# How to Be an Independent Video Producer

by Bob Jacobs

**Knowledge Industry Publications, Inc.**
**White Plains, NY and London**

*Video Bookshelf*

How to Be an Independent Video Producer

**Library of Congress Cataloging-in-Publication Data**
Jacobs, Robert M.
    How to be an independent video producer.

    Bibliography: p.
    Includes index.
    1. Video recordings--Production and direction.
2. Video tape industry--Management.   I. Title.
PN1992.95.J33  1986      791.45′0232      85-23702
ISBN 0-86729-180-X

Printed in the United States of America

10   9   8   7   6   5   4   3   2   1

## DEDICATION

This book is dedicated with love to two people: to Ray Bradbury for all the years of encouragement and free lunches at The Daisy when I really needed both; and to Martina Ann "Max" Romanowski, my partner in business and in life, for so much more than just the freedom and time to write this book.

## ACKNOWLEDGMENTS

The author wishes to thank the following individuals and organizations for their valuable help and support during the research and writing of this book: Apple Computer, Inc.; Claudia McCarley; Gary Cleveland; Martina Romanowski; Beverly Wiscinski; Frank Frassetto; Steven Gross; Video Trend Assoc.; Ronald J. Bullock; Fay McCarty; Tech Services, Inc.; Joseph Wiedenmeier; Metavision; Theo Mayer; Peter Inebnit; Ray Bradbury; Lee Wells; Dr. Robert L. Snyder; RAB; TvB; Michael O. Sajbel; John Jansen; Sam Drummy (and his two Emmys!); Robert D. Coglianese, CPA; The University of Wisconsin-Oshkosh; Dr. Stanley D. Sibley; Dr. William J. Leffin; Erik B. Ellingson, JD; Vincent Chieffo, JD; American Cinema Editors; Georgia Marcher; Berenice St. Bernard; Pasquale Ferrari; Norma Baker; and the producers interviewed in Hollywood who granted us generous amounts of their time.

And to Saul Bass and the memory of our beloved Verna Fields for their inspiration.

# Contents

# List of Tables and Figures

# Introduction

Producing is much more an art than a science; the producer works best as a creative member of a dynamic team, whether in entertainment, industry or education.

Ask ten non-producers just what a producer does and you'll probably get ten different answers. A typical director will tell you that the producer is the "business guy" who is largely concerned with budgets and schedules. A typical writer will say that the producer is the person who keeps insisting on changes in the script when the writer knows that it was fine the first time around. A typical editor will state that the producer is the one who has the final say in the "cut." All of them will be partially right, of course, but none will have told the entire story.

A quick definition of the producer's role might state that he or she is an organizer, a manager and a manipulator. The producer takes charge of a project, assembles all the necessary pieces or elements (including people) and sees the project through to completion.

What is lacking in that definition constitutes the substance of this book.

Defining the producer's job seems simple enough if we use the one-liners above. But consider how absurd the following role definitions appear in abbreviation:

- Video editor—pushes buttons to put together separate images.
- Director—moves actors around on the set.
- Cameraperson—points the camera and focuses it.

All of these are correct to a degree, but they are also misleading because they are so incomplete. We all know that these definitions are simplistic because of the enor-

1

mous amount of information that has been written and spoken about these artistic sides of the craft of film and video production.

Since we are living in an era of the director/auteur, the director accepts ultimate credit for a production and few laypersons question the validity of a single person laying claim to the entire work. However, as the end credits roll it becomes clear that without these large numbers of artists and technicians, the director would have no film on which to lay claim.

Any video production, just like any film project, requires a team, working in perfect harmony and with a single purpose. All a writer needs to write a novel is a pencil and some paper. All a fine artist needs to create a sculpture or a painting is materials and some tools. The art of video production is one of the few in the long history of creativity in which no single person, regardless of talent or intelligence, can go it alone. Whether the piece is a 30-second television commercial or a twelve-hour network mini-series, putting it all together takes a group of multifaceted individuals. Out of this group, the producer is the coordinator, the orchestrator, the single visionary and the captain of the creative ship.

As the person at the top, you will have to be all things to all people. You will have to assess and respond to the various needs of everyone involved. Everyone will expect you to be a pillar of strength, always consistent, always available, always ready to give them what they need. In the words of Rudyard Kipling, "If you can keep your head when all about you are losing theirs and blaming it on you," then you are a producer and are managing as the principles of producing dictate you must.

Since producers have the ultimate responsibilitiy for the outcome of any project, they also must know the basics of each of the technical and artistic functions of the group in order to understand what is and is not possible to achieve. Producers with no comprehension of elementary camera and lighting operations, for instance, will not be able to evaluate and budget properly for a project that needs a large number of camera setups.

Producers are also top management. As such, they must have a firm grasp of pure business details like proposals, budgets, profit and loss sheets, tax consequences, accounting, etc. This is especially true of independent producers who must take care of these details themselves. This book assumes that you know *how* to produce a videotape; it does not cover basic production techniques. Rather, it covers the business side of independent production.

All of the people who work for, or who one day hope to work for producers, must understand what producers are and what they are likely to expect from casts and crews. Another purpose of this book is to provide a guide through the maze of misunderstanding between producers and their staffs.

You will be assisted part of the way by advice and counsel from some of the most outstanding producers in the television world. These include Saul Turtletaub, Stephen

J. Cannell, Saul Bass, Henry Winkler and Gene Roddenberry. You'll also find quotes from some lesser known professionals who run small, profitable video production companies. These interviews, coupled with the other information presented in this book, serve to make *How to Be an Independent Video Producer* a useful primer on establishing an independent production business.

As the United States continues to become an information systems society, the opportunities for independent video producers are virtually limitless. In the following pages you'll find out why and how, with much determination, energy and a good bit of luck, you can take advantage of these opportunities.

This is a practical guide. For professionals in the field, we hope it will serve as a valuable refresher. For beginners in the business—those still in school or just about to venture out on their own—we hope to arm you with enough information to make your entry into independent production a smooth one, and to make your stay long, fulfilling and profitable.

Chapter 1 covers the business end of producing and Chapter 2 offers suggestions on managing people and productions. The reader learns how to set up the independent shop in Chapter 3 and receives some tips on promotion and marketing in Chapter 4. Chapter 5 discusses how to find and develop markets and Chapter 6 gives procedural and practical business hints.

# 1 The Business End of Producing

The independent producer, unlike someone who works for a large corporation, has one overriding duty: to sell the product in advance of producing it. Whether we call it "raising funds," "underwriting," "capitalizing" or any other euphemism, getting the money is what it's all about. And obtaining money involves the producer in the fine art of salesmanship. All independent producers should have taken at least a couple of college courses in basic business principles, especially marketing, accounting, management and sales.

## SELLING YOUR IDEA

Whether you are selling a dramatic series to NBC or a sales promotion tape to a widget manufacturer in your town, the process is essentially the same. Very simply, you must sell the idea. The six-step process that will help you accomplish this goal is as follows:

1. Assess the need
2. Write the proposal
3. Prepare a detailed budget breakdown
4. Show cost effectiveness to the client
5. Make an effective presentation to the client
6. Close the deal

We'll cover these steps one at a time, drawing on successful examples from the files of established producers. No one can guarantee that every idea a producer comes up with will sell. However, you can head the project in the right direction by getting organized at the outset and following these guidelines.

## Assessing the Need

Many people have the false notion that one success, one hit show, is all a producer needs to be given carte blanche from then on. They think that once your name is on a network show and you're rich and famous, all the doors in town are open. But as the line from *Porgy and Bess* goes, "It ain't necessarily so!"

Saul Turtletaub, who, with his partners Bernie Ornstein and Bud Yorkin, produced the successful show "Sanford and Son," sheds light on the truth of the matter: "The only thing a success does for a producer is to make the network executives willing to have a look at the next and the next idea. But if the new idea isn't a good one, if it doesn't fill a need of that executive, it doesn't matter what your name is or how many previous successes you've had. Every time I go in I have to be as well prepared as if it were my first shot out of the box."[1]

The word "need" as it applies to a network executive is fairly straightforward. A television show "needs" to have the most viewers possible in order to increase the network's Nielsen ratings. Unfortunately for all of us, no one has yet discovered the magic formula that will always work to attract the diverse and fickle American TV audience.

Assessing needs can be easier when you are trying to sell a commerical, a training tape or an educational video. When you are tackling anything other than a network TV show, your work should begin with some research about your prospective client.

Let's say that you're a producer in a medium-sized market (any community or group of communities in America with a population base of about 150,000). In that kind of market there are a number of large chain store operations such as fast-food restaurants, national hotels and motels, and home improvement stores.

You will also find a great many locally owned and operated businesses, and there will be some manufacturing. Most parts of the country have branches of major firms like Brunswick or Georgia-Pacific, along with smaller but no less viable local or regional companies. All of these are potential clients if you know what they need and how to give it to them.

### Discover Whom to Contact

The first thing you must do at each business is find out the name of the person in charge of the area in which you would like to make a sale. At larger corporations there will be titles like "Vice President: Marketing," "Manager: Training" or "Director: Corporate Communications." At smaller firms you're more likely to be dealing with an owner or a partner,

Whomever you talk to, keep this rule in mind: never deal with anyone who can't say yes. There is nothing more frustrating than making a wonderful pitch only to have the recipient say, "Gee, that was great. I'll have to tell my boss about it when he comes back from Bermuda." Many people within a company do not possess the power to say yes; therefore there is no point in wasting your time or theirs on a pitch.

So how do you find out the magic name of the "yes" person?

If it's a large corporation, send for a copy of the company's annual report to stockholders. These are usually slick and glossy publications, in which you'll find a list of all the corporate officers by title and the location of the company office at which they can be reached. Remember that if you want to sell a training video to a branch, for example, you frequently have to make your presentation to the decision makers at corporate headquarters. While affiliates and subsidiaries of major corporations exercise a degree of local control and autonomy, the budgets for media, advertising and training generally come from headquarters.

In the case of a smaller outfit, you can do a couple of simple things. You can visit the Chamber of Commerce or Organization of Manufacturers and Commerce in the community where the firm is located. Most chambers publish annual lists of their local businesses with the names, telephone numbers and addresses of the executives in each company. If your chamber doesn't provide this service, a friendly call to a receptionist will usually do the trick. The receptionist will usually provide the name of the person in charge and you can set up an appointment.

## Research the Client

Now that you know the name of the firm and the person, what are you going to say?

The typical slogan is "I have a video production house and I can meet all of your audiovisual needs." This slogan appears somewhere in the literature for many production companies—but it is far too general.

If you ask most businesspeople to tell you what they need, you're likely to be rebuffed. Therefore, it is imperative that you have at least one specific project to propose at your first meeting. Obviously, this means that you will have done some research into the nature of the business. You will know something of the jargon spoken, be familiar with the competition, be knowledgeable about the company's position in the field, and you should also know about its past experiences, if any, with video. An independent producer, therefore, must enjoy a continuing education in a wide variety of fields.

Let's say that you've chosen to propose a training video to a business with a typically high employee turnover rate—such as a hotel or motel. You believe you can demonstrate to the management that significant cost savings can be realized by training the housekeeping staff with a video aid.

Prepare charts and graphs based on the standard pay scale for supervisory time spent in training new personnel in such tasks as making beds, cleaning, taking inventory, stocking, and so forth. You can assume that each hotel or motel also spends some time discussing grooming, uniforms, etc. Explain how your visual aid will save the hotel time or money.

You can make your case for a visual training aid on the grounds that they have been demonstrated to cut the time it takes to learn. (Give examples such as the U.S. Government implementing training films during World War II and drafting such luminaries as John Ford and Frank Capra to make them.) Even people who can't or don't like to read can respond to a movie or a video that entertains while it teaches.

In order to present a good proposal, you must be able to speak authoritatively about the hotel/motel business. What do you do, short of taking a crash course in hotel/motel management? Contact a few of the many sources available that can help with your research.

Most industries have associations. In many cases you don't have access to the inner trade secrets of those associations unless you're a member. It doesn't hurt to try, though.[2]

Two other major sources of information for producers are the Radio Advertising Bureau (RAB) and the Television Advertising Bureau (TvB). Both organizations provide research services to their members. The fees vary depending on the market size of your area, but the access to information is worth almost any price.

Figures 1.1 and 1.2 are reprints of the RAB and TvB "Instant Background" profiles for hotels and motels. Look them over carefully; as long as you have this type of information, you are well prepared to make your initial call to any hotel/motel manager.

When you know and can discuss times of peak business, sources of revenue, how the bills are paid, and consumer attitudes by a demographic breakdown, you will sound authoritative in the hotel/motel business. Without being overbearing or trying to tell clients that you know more than they do, you can gain clients' confidence as a colleague.

Both the RAB and the TvB have profile information on every major kind of business in the United States, from shoe stores to heavy manufacturing concerns. These are invaluable resources for the independent producer. Membership in one or both bureaus also brings monthly tips on sales techniques and access to commercial spots that you can adapt to your needs.

### Keep Informed about Related Fields

As you can see, the good producer must be a veritable storehouse of information. One should read virtually everything available to keep up to date on technological and

**Figure 1.1: Radio Advertising Bureau Profile for Hotels and Motels**

RAB INSTANT
BACKGROUND:

# Hotels & Motels

Size of the Business. The U.S. Department of Commerce gives total receipts of the lodging industry:

| | | | |
|---|---|---|---|
| 1981 | $26.637 billion | 1979 | $22.226 billion |
| 1980 | 23.638 billion | 1978 | 19.443 billion |

There are about 2,525,000 rooms available today, of which the top 25 chains represent 45% (Travel Market Yearbook).

When Business Occurs. Percent of receipts by months (U.S. Dep't. of Commerce, 3-year average):

| | | | | | |
|---|---|---|---|---|---|
| January | 7.1% | May | 8.6% | September | 8.5% |
| February | 7.3 | June | 8.9 | October | 8.8 |
| March | 8.5 | July | 9.5 | November | 7.5 |
| April | 8.2 | August | 9.8 | December | 7.2 |

Occupancy Rates. For several years 67-70% of all rooms have been in use on the average day. By location, type, and size (Laventhol & Horwath):

| Location | | Region | | Size (No. of Rooms) | |
|---|---|---|---|---|---|
| Airport | 70.1% | Northeast | 71.6% | Under 150 | 68.3% |
| Suburban | 68.0 | N. Central | 61.2 | 150-299 | 66.6 |
| Highway | 67.2 | South | 66.8 | 300-600 | 66.5 |
| Downtown | 65.9 | West | 71.3 | Over 600 | 67.2 |

By Type and Incidence of Double Occupancy (Pannell, Kerr, Forster):

| Type | Occupancy | Double Occupancy As Percent Of Total Occupancy |
|---|---|---|
| All places | 67.6% | 48.4% |
| Transient Hotels | 66.9 | 40.0 |
| Resort Hotels | 68.4 | 87.5 |
| Motels with Restaurants | 68.1 | 43.2 |
| Motels without Restaurants | 70.2 | 45.3 |

Sources of Revenue. The Panell, Kerr, Forster study shows:

| | All Lodgings | Transient Hotels | Resort Hotels | Motels With Restaurant | No Restaurant |
|---|---|---|---|---|---|
| Rooms | 60.2% | 60.4% | 53.1% | 64.0% | 93.1% |
| Food | 23.8 | 24.3 | 26.6 | 21.6 | ---- |
| Beverages | 9.0 | 9.1 | 9.0 | 9.9 | ---- |
| All other | 7.0 | 6.2 | 11.3 | 4.5 | 6.9 |

Price Ranges. 12% of guests stay in deluxe/first class accomodations, 75% average/middle price, 13% economy or budget. This is the same for both business and pleasure travelers (Travel Market Yearbook).

Reproduced with the permission of the Radio Advertising Bureau.

**Figure 1.1: Radio Advertising Bureau Profile for Hotels and Motels (Cont.)**

Handling of Reservations. 10% of a travel agent's business comes from lodging reservations, and 69% of agents have automated systems.  59% of reservations placed by agents are guaranteed ( of them, 57% covered by client credit cards, 29% prepaid vouchers, 14% the agent's own credit) (Travel Weekly Harris Poll).  Incidence of reservations to total hotel guests, and who places them (Laventhol & Horwath):

|  | Downtown | Airport | Suburban | Highway | Resort |
|---|---|---|---|---|---|
| Percent with Reservations | 81.9% | 66.6% | 62.1% | 67.0% | 88.4% |
| How Placed: Direct | 38.0% | 35.2% | 41.5% | 46.4% | 42.0% |
| Reservation Systems | 39.4 | 46.3 | 49.2 | 36.9 | 17.7 |
| Travel Agents | 16.0 | 12.2 | 8.1 | 10.8 | 24.4 |
| Other | 6.6 | 6.3 | 1.2 | 5.9 | 15.9 |

A Time survey of hotel guests found 61% of nonbusiness travelers made reservations more than one month in advance, but only 8% of business travelers did.

Decision-Making. 71% of travel agent vacation customers are given advice on choosing hotels; same for 41% of business customers (Travel Weekly).  The Time survey found the guest chose which hotel to stay in 77% of the time, someone else in the company 31%, travel agent 4%, airline 2% (more than one influence could be listed, also more than one hotel could have been used on the same trip).  The most important factors influencing choice were prior experience 25%, location 24%, recommendation from business associates 24%, image 8%, cost 7%, travel agent 3% (others unimportant).
Percent of guests considering these facilities "very important":

|  |  |  |  |
|---|---|---|---|
| Beds | 84% | Restaurants | 66% |
| Friendly staff | 77 | Towels | 65 |
| Bathroom | 72 | Housekeeping services | 64 |
| Reservation service | 70 | Value for price | 59 |
| Safety, security | 69 | Wake-up call | 57 |

Meetings & Conventions. 92% of attendees stay at hotels/motels (Travel Market Yearbook).  Percent of planners using (at various times):

|  | Associations | | Corporations |
|---|---|---|---|
|  | Major Conventions | Seminars | Company Meetings |
| Downtown Hotels/Motels | 61% | 56% | 54% |
| Resort Hotels/Motels | 36 | 38 | 61 |
| Suburban Hotels/Motels | 16 | 52 | 55 |
| Airport Hotels/Motels | 9 | 46 | 37 |

Profile of Guests. Half are repeat trade; 90% U.S. residents/10% foreign (Laventhol & Horwath).
Distribution of lodging expenditures (U.S. Travel Data Center):

| Purpose |  | Age of Travel Party Head |  | HH Income |  |
|---|---|---|---|---|---|
| Business/convention | 29.0% | 18-24 | 11.1% | $35,000+ | 14.3% |
| Visit friends, | 10.4 | 25-34 | 19.1 | $25-34,999 | 17.7 |
| relatives |  | 35-44 | 21.3 | $20-24,999 | 14.0 |
| Personal business | 8.8 | 45-54 | 18.5 | $15-19,999 | 17.2 |
| Outdoor recreation | 21.7 | 55-64 | 12.7 | $10-14,999 | 16.6 |
| Other tourism | 17.7 | 65+ | 17.3 | Under $10,000 | 20.2 |
| Other reasons | 12.4 |  |  |  |  |

| Vacation | 52.9% | Number on Trip : | 1 | 48.4% | 3 | 6.3% |
|---|---|---|---|---|---|---|
| Not a vacation | 47.1 |  | 2 | 31.7 | 4+ | 13.6 |

**Figure 1.2: Television Advertising Bureau Profile for a Specific Hotel**

CATEGORY:      Hotels & Resorts
MARKET:          Baton Rouge
ADVERTISER:   Sheraton Hotel

The Sheraton Hotel in suburban Baton Rouge used television to expand the image of the hotel. Originally perceived as "just a hotel" by the local residents, the Sheraton has become a major meeting center for business and social activities.

The successful re-positioning of the hotel was accomplished by using two types of commercials. One promotes the various facilities of the hotel (dining room, meeting rooms) and special weekend packages. The other commercial is used to promote live entertainment in the lounge and is changed when a new act is appearing.

The Sheraton advertises on the local news and buys various specials, allocating 70% of its total advertising budget to television.

According to Jayne Rule, the hotel manager, "television has increased our sales and fulfilled our marketing objectives."

Source:    WBRZ-TV, Baton Rouge, August 1979

B21

Television Bureau of Advertising, Inc., 1345 Avenue of the Americas, New York, NY 10105

Reproduced with the permission of the Television Advertising Bureau.

business trends, politics, social change, etc. Many producers subscribe to a number of magazines and newspapers to keep up to date, the bulk of which are tax deductible as professional subscriptions. If you assimilate as much information as you possibly can, you will be amazed at how often the knowledge triggers good, salable ideas, or comes to your aid in a presentation to a client.

Many people these days claim they don't have time to read. Nonsense! Take an hour a day. Close your office door and skim articles. Put a stack of literature next to your bed and read every night before you go to sleep. Read over breakfast and lunch, and in snatches as you go through your day. After all, there are 24 hours in each day and you can certainly find between 60 and 120 minutes to do your homework.

Here is a minimum list of suggested periodicals (aside from video trade journals) that will fill your gray matter with great, usable information: *Science '86; Science Digest; Discover; Psychology Today; Time; Newsweek; U.S. News and World Report; The Wall Street Journal* (the best source of information about the business of business); *The New York Times; The Washington Post; Changing Times; The Hollywood Reporter; Variety;* and *Rolling Stone.*

When it comes to trends, a producer should be at the leading edge in order to anticipate needs, forecast future developments and speak knowledgeably with both clients and colleagues. Adding many of the aforementioned titles (you can choose one or two of the news magazines) to your reading list will help keep you on that edge. Add other titles as your own needs dictate. For example, if you want to break into the lucrative horseshow trade, subscribe to such magazines as *Western Horseman* and *Horse and Rider.* If you want to clinch a deal with a boat manufacturer, you will find *Yachting, Sea and Pacific Motorboat* and *Lakeland Boating* useful in preparing your presentation.

To find specific material, no matter how esoteric, go to your local library. *The Reader's Guide to Periodical Literature* lists by subject every article published in magazines, newspapers and journals. There is a growing number of computer services throughout the country that provide users with access to a wide variety of specific research topics. If you have a personal computer and a telephone modem, you can join one or all of them and conduct in-depth research from the comfort of your own desk. Check with your local computer dealer to find out what is available in your area or subscribe to one or more of the computer magazines that serve this field and which advertise computerized research networks and services.

Every time you make a sales call, think of it as a battle of wits. Solid information and your own personality are the only weapons you have. Make sure that you are armed with both when you begin.

## Meet with the Client

After you have gained both access to, and the confidence of clients, it is time to let them explain the needs of the company. What you are actually doing is leading the client down the path to a project that you have already conceived. In the case we are using here, you are trying to sell a training tape.

Devise a questionnaire like the one illustrated in Figure 1.3. Use it as a consulting form so that the client opens up to you. Use a carbon sheet so you can leave a copy of it with the client when you leave. This informational meeting is, of course, offered to the client with no obligation. You hope, of course, that by the time you have finished your six-step procedure, you will have locked up a contract!

Accentuate the positive as you fill out the questionnaire. Keep the client answering your questions in the affirmative by phrasing statements like, "It is true that you have a high turnover rate with your housekeeping staff, isn't it?" Follow with, "And I'm sure you'd like to be able to reduce the costs of training new people, wouldn't you?" You can count on the response to that one from almost anyone!

By the time you reach the end of your presentation, the client is so used to saying yes to you that in a majority of cases you'll also get the desired response when you drop the bomb, "Since we are in total agreement on your need for a high-quality training tape, there is no reason we shouldn't sign a contract and get it done." There will be more about this aspect of the sales job when we discuss closing the deal later in this chapter.

At this point, thanks to your research, your personable and timely fact-finding mission with the client, the information you have garnered from the questionnaire and a solid feel for the potential amount of money this client has to spend on your project, it's time to get to the office for the next step.

## Writing the Proposal

There are no pat formulas for proposal writing. The best advice runs along the lines of what high school teachers still tell students about doing a term paper:

1. Tell 'em what you're gonna tell 'em
2. Tell 'em
3. Tell 'em what you told 'em

In Appendix E you'll find two sample proposals, one for a motorcycle company, and another for a major manufacturer of outboard motors. Both are representative of standard proposals.

The key thing to remember is: be concise and to the point. You are not writing a book, a technical article or a script. Use many action words; write as if doing the project is a foregone conclusion. At this stage what you are really doing is putting on paper what has already been discussed in your first meeting with the client. The following is a general outline that you can use until you develop an individual style of your own.

**Figure 1.3: Questionnaire to Evaluate Media Needs**

## Account Consultancy Questionnaire

**To the Client:** This consultancy form is a confidential document used by us to evaluate your media needs. The information that you give us will not be shared with anyone else. Your account executive will leave one copy of this form with you at the conclusion of the interview. There is no obligation to you for this consultancy. It is provided free of charge. You may use any of the results from it for any purpose you choose. We will use the results to formulate a proposal in writing to provide media services to you.

## Company Background

1. Name of Firm: _____

2. Type of business: _____

3. Number of years in business: _____

4. Company Officers:

| Name | Title | Duties | Phone |
|------|-------|--------|-------|
| | | | |
| | | | |
| | | | |

5. Address:_____

6. Annual Gross Income: (optional) $ _____

7. Number of employees: _____

8. Who are your primary competitors in this area?_____
_____

9. What percentage of your business do you estimate the competition is taking away? _____
_____

## Assessing Your Needs

1. What are your company's strengths? _____
_____

2. What are your company's weaknesses?_____
_____

3. What would you personally like to change? _____
_____

4. What additional share of the market would you realistically like to obtain this year? _____

5. In what areas would you most like to reduce operating costs? _____

6. What are your operating costs in the following areas?

       [a] Training . . . . . . . . . . . . . . . . $ _____

       [b] Advertising . . . . . . . . . . . . . $ _____

       [c] Marketing . . . . . . . . . . . . . . $ _____

       [d] Sales Promotion . . . . . . . . . $ _____

       [e] Safety . . . . . . . . . . . . . . . . . $ _____

## Figure 1.3: Questionnaire to Evaluate Media Needs (Cont.)

7. When does most of your business occur?_____

8. How many locations do you have? _____

9. Where are they?_____

_____

10. Do you have demographic figures on your target market?

   [a] Yes      [b] No

11. Would you like us to provide them at no cost to you?

   [a] Yes      [b] No

12. What is the media mix for your advertising?

   [a]_____% radio    [b]_____% TV    [c]_____% newspaper

   [d]_____% magazine    [e]_____% other _____

13. Do you employ an advertising agency?

   [a] Yes    [b] No

14. If so, name, address and contact:_____

_____

15. What was your most successful ad campaign? _____

_____

16. When was it? _____

17. How did you measure its success? _____

_____

18. What was your least successful ad campaign? _____

_____

19. When was it? _____

20. How did you measure its failure? _____

_____

21. What is your employee turnover rate?_____

22. How are new employees trained?_____

_____

23. What is the major strength of your present training program? _____

24. What is its major weakness? _____

25. What major obstacles impede your sales force? _____

_____

_____

_____

26. Do you perceive that a motivational or sales promotional video would help remove one or more of these obstacles?

   [a] Yes      [b] No

27. Have you seen one or more such videotapes?

   [a] Yes      [b] No

28. May we make a presentation of our ideas to you and your staff at no obligation to you?

   [a] Yes      [b] No

29. What is a convenient date and time for the presentation? _____

## Title Page

Print the title page on your letterhead. A little less than halfway down the page, write the title of the project you have in mind. About three-quarters of the way down the page, on the left margin, list to whom it is presented, who wrote it and when.

## Page One

On no more than one page, state your case. Tell specifically what you propose to do. Your first paragraph to the hotel manager will say something like:

> We propose to do a training videotape about housekeeping tasks. The purpose of the tape is to demonstrate all aspects of these tasks, focusing on the unique approaches that make Jones Hotels the hallmark of good service. The tape will be used by your training staff to reduce the amount of time spent in this high-turnover area of your business and, because of the proven effectiveness of audiovisual training aids, to help make your housekeepers better and more efficient then ever.

Flesh out the rest of the page with the data you have collected in your research and on which you and the client have already agreed. Once more, try to keep the client thinking in the affirmative. Do not, at this stage of the game, come up with new information or some outlandish idea that occurs to you while you're writing. It is imperative that this written proposal be the essence of the contractual agreement that you will make at the end of the process. Lay the foundation for your project and then leave it alone.

## Pages Two and Three

On these pages, lay out in very specific terms what your project is going to be. Get to the point quickly and be concise. By this time you will have outlined the script on paper; from this outline, write a treatment—a description of your script in paragraph form. In our example, the first portion of your treatment might look something like this:

> Our production opens with an introduction by Harold Jones, president of Jones Hotels. Mr. Jones will explain that the housekeeping staff is the key to the entire operation, since the reputation of the hotel is one of personal service. We find that having top management appear in this capacity gives employees a good feeling that they are cared about and not treated as peons.
>
> Following Mr. Jones's introduction, we will feature Harriet Smith, director of housekeeping, who will explain that the tape will teach new staff members the inner secrets that have made Jones Hotels such a desirable stopover for guests here in _____ City. She will show a chart listing all the areas to be covered in the production:

- Bedmaking
- Room cleaning
- Flower arranging
- Uniform appearance
- Laundry services
- Inventory

After this brief introduction we will move through each of the above steps, using professional performers to portray housekeepers. We will work on location at a Jones Hotel for authenticity. Each training segment will be no more than five minutes long, with cutbacks to Ms. Smith for quick review between segments. We will provide printed review forms that ask questions taken from the text of the video presentation. These may be retained by the trainees and used for refreshers on the job.

Depending on the length of the project, this section of your proposal will take about two pages. Under no circumstances should it take more than three pages to tell your story; otherwise your reader may lose interest. As Shakespeare had Polonius say, "Since brevity is the soul of wit, I will be brief!"

## Page Four

Here you might wish to include a sample page of the script. Laypersons are frequently impressed by seeing the script form; it makes them feel like insiders in a business that has glamour and excitement. (Don't tell them that's why you're putting it in.) Explain that you have included a sample page of the script to show them how it is going to look on paper.

If you elect not to show a sample page of the script, this page can be used to further discuss your treatment, explaining to the client that you are fully prepared to begin writing the script upon contract approval.

## Page Five

In order to make a decision, any businessperson is going to have to see the "bottom line." And this is where you present it in the form of a Production Budget Summary. Figure 1.4 is a reproduction of a standard form used throughout the industry. Several of the categories listed on this form do not apply to smaller productions. You can make up a form using as much of this one as you like. Have it printed in order to look as professional as possible. You arrive at the summary figures through a more complex budgeting process which we will examine in detail later in this chapter.[3] The detailed figures are generally not shown to the client.

## Page Six

This is your conclusion page. Use it to "tell 'em what you told 'em." You should always end on a positive note, like this:

**Figure 1.4: Sample Production Budget Summary***

# Budget Summary 8126

PROD. TITLE _____ NO. _____ DATE PREPARED _____

PRODUCER _____ DIRECTOR _____ ESTIMATOR _____

PROD. START DATE _____ ESTD. FINISH DATE _____ TOTAL PROD. DAYS _____

| ACCT. NO. | DESCRIPTION | PAGE NO. | BUDGET | COST TO DATE | REMARKS |
|---|---|---|---|---|---|
| 100 | STORY & OTHER RIGHTS | | | | |
| 110 | WRITING & RESEARCH | | | | |
| 120 | PRODUCER & STAFF | | | | |
| 130 | DIRECTOR & STAFF | | | | |
| 140 | TALENT | | | | |
| | | | | | |
| 190 | FRINGE BENEFITS | | | | |
| | ABOVE THE LINE TOTAL | | | | |
| 200 | PRODUCTION STAFF | | | | |
| 210 | CAMERA | | | | |
| 220 | ART DEPARTMENT | | | | |
| 230 | SET CONSTRUCTION | | | | |
| 240 | SET DRESSING | | | | |
| 250 | SET OPERATIONS | | | | |
| 270 | ELECTRICAL | | | | |
| 280 | PROPERTY | | | | |
| 300 | SPECIAL PHOTOGRAPHY | | | | |
| 320 | SPECIAL EFFECTS | | | | |
| 330 | WARDROBE | | | | |
| 350 | MAKEUP & HAIRDRESSING | | | | |
| 360 | SOUND (PROD.) | | | | |
| 380 | LOCATION EXPENSES | | | | |
| 400 | TRANSPORTATION | | | | |
| 410 | FILM & LABORATORY (PROD.) | | | | |
| 420 | FILM TESTS & SUNDRY | | | | |
| | PRODUCTION TOTAL | | | | |
| 500 | EDITING & PROJECTION | | | | |
| 510 | MUSIC | | | | |
| 530 | POST PROD. SOUND | | | | |
| 550 | POST PROD. FILM & LAB | | | | |
| 570 | TITLES, OPTICALS & INSERTS | | | | |
| | EDITING PERIOD TOTAL | | | | |
| 600 | INSURANCE & TAXES | | | | |
| 610 | PUBLICITY & STILLMAN | | | | |
| 620 | FRINGE BENEFITS (below the line) | | | | |
| 630 | OFFICE EXPENSE & LEGAL FEES | | | | |
| | OTHER COSTS TOTAL | | | | |
| | DIRECT COST TOTAL | | | | |
| 700 | INDIRECT COSTS | | | | |
| 720 | COMPLETION BOND | | | | |
| | GRAND TOTAL | | | | |

DON KIRK ENTERTAINMENT ENTERPRISES   RT 9, BOX 127, CANYON LAKE, TEXAS 78130

Reproduced with the permission of Don Kirk Entertainment Enterprises.

*Some of the production forms reproduced in this book are from a collection available as a package from Don Kirk Entertainment Enterprises, 1021 W. Mulberry, San Antonio, TX 78201.

We are very excited about working on this project with Jones Hotels. We know that your housekeeping staff will enjoy our presentation and that you will enjoy the considerable cost savings it will bring to your operation. We look forward to a long and happy association with you.

## General Tips

Finally, here are a few tips about proposal writing that you should keep close at hand:

1. Remember that this document is what your client is going to keep after you have gone. It is a primary selling tool. You are going to be held to whatever you say in this document when it comes time to deliver the final product. Make sure you can deliver what you say you can.
2. Use positive language. Do not use phrases like, "We think" or "We believe." Instead, say, "We will," "We do," "We are," etc. Any negativity that appears in the language or style of your proposal will exert a subtle but real effect on your clients as they pore over the document.
3. While being as positive as possible, avoid hyperbole. Don't blow your own horn. This does not mean that you should engage in false modesty. The late actor Walter Brennan used to say it quite well as a character in "The Guns of Will Sonnett": "No brag. Just fact."
4. This tip may seem obvious, but it needs to be stressed. Proposals should always be typed on a high-quality typewriter, double spaced, with page numbers at the top. If your only typewriter is an antique Remington handed down to you by Grandpa, then engage the services of a professional typist. Better still, take the manuscript to a printer and have it set, and duplicate enough copies to that your client can pass them around to colleagues.

   A few dollars spent on making your printed work look top-notch will pay dividends in the end. If you are one of the millions who have moved into computerized word processing, never submit anything printed by the dot matrix method. Even the so-called "near-letter quality" printers cannot do justice to a proposal.
5. Always submit your proposal in an attractive binder. A three-ring plastic binder will do, but most printers and many photocopy service stores offer even snappier covers. You might even want to have a number of proposal covers printed with your company name and logo on the front for a classy, professional look. Let your good taste—and your pocketbook—be your final guide.

## Preparing a Detailed Budget

Budgeting a project is probably the toughest, most time-consuming part of producing. After all, it dictates your bottom line. Budgeting requires care and a keen analytical mind. If the production is underbudgeted, you risk losing your shirt. If it is overbudgeted, you risk losing the client. Walking the balance beam between these two is the heart of successful producing.

Budgeting is not a mysterious process; there are some sensible and time-tested techniques to assist you. It involves consideration of everything you can possibly think of. A set of Picture Budget Detail forms that the industry has developed for just this task is reproduced in the Appendix to Chapter 1 (see page 35). Figure 1.4, the Production Budget Summary, is the cover sheet for this rather lengthy and detailed set of forms that will lead you through the budgetary maze. These forms were originally conceived by the motion picture industry. With the exception of category 12, "Cutting-Film-Laboratory," the forms apply equally well to video production.

Items 1 through 7 are called "above-the-line" costs. Items 8 through 22 are "below-the-line" costs. This arbitrary "line" separates creative elements such as talent and script from the technical aspects of production.

As you read through these forms you will notice that each category from the summary sheet is broken out into a number of subheadings and line items. In the case of our hotel training tape, it is quite unlikely that we would need all of the personnel listed in Category 9-C (Wardrobe Dept.). The process of considering all of the possibilities in this manner will, however, sharpen your mind and bring out many things that might otherwise have been overlooked, only to surface later, when it would affect your bottom line. And by following the budgeting steps outlined here, when you do get ready for that mini-series or network television special, you'll be fully prepared to deal with a complex budget.

## Estimating Time

In looking at the budget forms, you'll see that they indicate a column for days, weeks or quantity. Next to that is a column for the rate per day, week or quantity. All you have to do is figure out how many days or weeks you will be needing your staff, supplies and equipment and how much each category is going to cost. Add in a profit margin and you have a budget total.

The process can be just as simple as it sounds. All it takes is a logical approach to the situation and a means for determining worth.

Step one in budgeting is figuring out the time needed for each category. The Script Breakdown form shown in Figure 1.5 is a standard form that you can reproduce and use for this purpose. It is a budgeting tool with which you go through the script page by page and scene by scene, and project the needs for the shoot. Using this tool, you can project to the quarter hour who and what you will need and where and when you'll need them. If you had a script at this point, the Script Breakdown would be your primary tool.

You do *not* have a script yet, however. The reason you do not is simple; you have not been paid to write it because you haven't sold the project. From time to time you will present a script on spec hoping to sell it. For now, we'll assume this is not the case.

**Figure 1.5: Script Breakdown Form**

# Script Breakdown 8138

| PRODUCTION TITLE | | | SCRIPT DATED | PAGE NO. |
|---|---|---|---|---|
| SET | | SEQUENCE | | |
| PERIOD | | SEASON | DAY | NITE | INT | EXT | TOTAL SCRIPT PAGES |

| CAST | BITS | SCENE NUMBERS & SYNOPSIS |
|---|---|---|
| | **EXTRAS** | |
| **PROCESS / EFFECTS / CONSTRUCTION** | | |
| **MUSIC / MISCELLANEOUS** | | |
| **PROPS / ACTION PROPS / ANIMALS** | | |

DON KIRK ENTERTAINMENT ENTERPRISES   RT 9, BOX 127, CANYON LAKE, TEXAS  78130

Reproduced with the permission of Don Kirk Entertainment Enterprises.

You do, though, have a very good notion of what the script will be like when you do get around to writing it. Gene Roddenberry, creator and producer of the "Star Trek" television series says, "Producers in television tend to be writers or writer types. Producing is just an extension of the storytelling process."[4] So we'll assume that you are going to be the writer. Without actually writing the script, you can draw on your research, your visit to the hotel and your discussion with the client to do a quick outline of the project. From this outline you can draw some conclusions that will allow you to use the Script Breakdown for budget estimating.

To predict the number of days in the shoot (discussed further in Chapter 2), you must first have an idea of how many camera setups will be involved. Each time the camera and lights have to be moved, time is consumed. A good rule of thumb is 30 minutes per setup once the master scene has been lit. In this stage of planning, overestimate the number of setups unless you are very experienced or are so familiar with your crew that you know how fast the changes can be made. From this estimate, you'll have a fairly accurate idea of the number of hours to be broken down into days and entered on your Shooting Schedule. A typical shooting day is ten hours.

## Estimating Equipment and Extras

The next factor to include in your budget is the cost of all the equipment you will need. If it is a relatively small shoot, you may not need much more than a camera with a variable focal length lens, a videocassette recorder, a microphone and a three- or four-unit lighting kit. When using nonprofessional talent, even on a small shoot, you'll find that the addition of a second camera and operator will save time and trouble in post-production. Pros, of course, can repeat actions interminably for matching cuts in post-production. Amateurs almost never can. Multiple camera use gives you real-time cuts on the action of the best performances.

Whatever equipment you will need must be spelled out thoroughly here. Some producers own the basics—a camera, VCR, light kit and so forth. Many do not simply because of the enormous investment involved. That is why rental equipment houses have spread from the major production centers like Los Angeles and New York.

Rental houses are now located throughout the country.[5] All have free catalogs that you can obtain simply by calling or writing for them. Most houses have daily and weekly rates; the weekly rate is typically either three or four times the daily rate. All of the rental firms ship anywhere in the country. You will have to know how much they charge for shipping as well as whether they charge you for the time the equipment is in transit. In addition to equipment considerations, you must attempt to project all of the *things* you will need to complete the production. This stage requires careful thought and attention to minute detail. These things include props, costumes, makeup, shipping boxes—everything that you, your cast and your crew will have to use. Budgeting forms are very helpful at this stage in organizing your thinking.

## Budgeting for People

Now you are ready to consider the next major budget item: people. Two categories of technicians and performers are available as cast and crew: union and nonunion.

If your project is being prepared for national television broadcast, you have no choice but to hire a union cast and crew. There are two major unions for technicians: NABET (National Association of Broadcast Employees and Technicians) and IATSE (International Alliance of Theatrical Stage Employees).

Video performers also have two unions or guilds. They are SAG (Screen Actors Guild) and AFTRA (American Federation of Television and Radio Artists). The addresses and telephone numbers of these four unions are listed in Appendix A.

Major producers and production companies are signatories to the union or guild contracts and, as such, are committed to hiring *only* union or guild members. Violators can be penalized with fines and production halts.

If you are a signatory producer, the job of budgeting for your cast and crew is simplified. Each guild has a set of minimum pay scales, including fringe benefits and payment schedules for repeat runs of the work, called residuals. Any signatory can obtain copies of the guild or union rules, regulations and fee schedules simply by writing for them.

The unionization of the production industry is a controversial issue. Many people feel that the unions have acquired too much power and set rates so high that they have driven producers out of major centers of production. Others feel that the unions and guilds provide a valuable service by establishing a system of internal screening and testing of candidates for admission to the guild or union, thereby guaranteeing that there will be highly qualified technicians in each category. If you are producing a show for NBC, you have no choice in the matter. For most other applications, such as our training tape for Jones Hotels, you most certainly have a choice.

Figuring the worth of nonunion personnel is a matter between you and them. A typical free-lance rate for a competent videographer is $300 to $500 per day. Frequently you can hire operators with their own cameras and VCRs for that amount. Remember that this person will be more responsible than anyone else for the final look of your piece, and therefore it may be foolish to scrimp on a fee.

The number of crew people you will need is entirely dependent upon the size and scope of the project. Unless you are someone who can, and wants to, do some of the technical work yourself, a typical minimum crew (in addition to the producer) for a small-to-moderate production consists of a director, videographer, VCR operator/sound recordist, camera assistant/gaffer, grip and script supervisor.

If you are not familiar with all of these terms, some brief descriptions follow. A gaffer is an electrician. This is the person responsible for setting the lights, patching into power sources if necessary, and so on. Many videographers have long-standing relationships with gaffers who also double as camera assistants to pull focus, push dollies, etc., in nonunion settings.

A grip is a general helper who will load and unload equipment, assist the gaffer, arrange the set, etc. Working without at least one grip on the set puts a major strain on everyone involved.

The script supervisor is responsible for continuity, for noting changes and assuring that the script is being shot as it was written. Script supervisors will also time shots with a stopwatch so at the end of each day you will know about how long your project is turning out to be. Their notes are used in post-production by you and the editor.

As your production develops, you will have to add additional categories of people from the Picture Budget Detail form. For the Jones Hotel shoot, we will assume that you can handle makeup, wardrobe and so on yourself. However, if you don't feel competent in these areas, employ someone who is.

When you have figured the amount for cast and crew salaries, be sure to add on the following:

- Meals (at least two a day, with one hot)
- Coffee, juice, soft drinks on the set
- Travel expenses
- Lodging, if necessary
- Liability insurance for all personnel and equipment

As an independent producer, in most cases, you will not have to compute withholding income taxes, workmen's compensation insurance or social security taxes because your free-lancers do not work for your company. You employ people on a per-project contractual basis as independent craftspeople who are solely responsible for reporting their own wages to the Internal Revenue Service. Chapters 3 and 6 discuss these and contractual considerations in more detail.

## Budgeting for Yourself

When you have arrived at cost figures on a daily or weekly rate for the crew and cast, direct your attention to yourself. The producer is, of course, entitled to a fee for each production over and above the profit margin that you will plug in for your company. Many novice producers forget this important factor, or figure that they will pay themselves out of the profit. If you are going to stay in business, remember that the business itself needs to be paid on each project. This profit is what will sustain the business and you through periods during which you have no work.

There is no fixed formula for determining a fair profit margin for the company. Some producers, because they do a large volume of business, can operate on a profit of as little as 5% of the total of each production budget. Others, if the client can afford it, have a built-in profit factor of 50% of the total production budget. Profit ultimately depends upon your needs, your client's pocketbook and your conscience.

After estimating time and equipment, and budgeting for staff and profit, it is time to move to the second phase of preparing a detailed budget—projecting finishing costs and unexpected costs.

### Finishing Costs

Finishing costs are those for post-production and completion. You will have to estimate how many minutes or hours of tape you will end up with, and approximately what the shooting ratio will be. A typical shooting ratio for a tape similar to the Jones Hotel tape is 5:1. This means you will shoot five minutes of tape for every one that makes it to the final product. If you have projected a 15-minute finished tape, you'll be editing it down from 75 minutes of raw footage.

You will do your first edit offline. This is the stage in which you make a rough-cut and log the footage numbers where the final cuts will take place. The final edit at an online facility is very expensive. For example, you can find offline editing at fees ranging from $25 to $100 per hour, whereas online fees run from $250 to more than $2000 per hour. Estimating post-production time is, therefore, critically important.[6]

Music is your final post-production consideration. There are a number of stock music libraries that provide music which you pay for on a needle drop basis.[7] Original music is preferable, of course, because it can be composed and performed to meet the specific needs of your production. There is a growing number of audio production studios that specialize in original scoring (see Chapter 5). Whether you use stock music or original scores will probably depend on your budget limitations, but either type of music can contribute a vital professional polish to any video production.

### Budgeting for the Unexpected

When you have tabulated all the costs on your budget form, you come to the block marked "Contingency." All producers will have undoubtedly overlooked some small detail in the budgeting process. Even if you haven't, an unforeseen emergency might arise during the production. Your lead actor could get sick and miss a day or two of shooting. Your videographer could quit and have to be replaced. Your shoot could be rained out for days on end. These are some of the reasons for the contingency fund. It is a built-in "fudge factor" that insures the project against the unexpected. The typical contingency runs between 20% and 40% of your total budget. It is tallied in at this point to give you the grand total.

When you have completed this process, you have done everything possible to estimate the real production cost. Use the completed budget summary form through each phase of the production, striving at all times to bring the project in at no more than the figures slated for each category and, if you are good and lucky, under budget. There is nothing in the world a client appreciates more than a producer returning, along with the final tape, a check for the amount a project came in under budget.

## Demonstrating Cost Effectiveness

At your presentation meeting with clients, lay the budget summary before them. For a project like the Jones Hotel training tape, the figure might typically run between $8000 and $15,000. In many cases this sum may be a considerable saving compared to the annual costs of training programs in wages and time. If so, you are well on your way to wrapping up the deal.

In other cases the cost of the program may be more than the company is already spending on training. Your job here will be to convince the client that the effectiveness of the training aid will enhance the quality of training. It might, for example, make the trainer's job simpler because the trainees will have access to the tape for refreshing on their own. Perhaps the fact that Mr. Jones has provided his staff with this state-of-the-art training aid will make his people think more highly of him, of themselves and of the Jones Hotel.

Since you know that your product is top quality, and that it will help the clients and their businesses in spite of the cost, you can resort to the emotional sales technique with a clear conscience. If you can't justify the product in your own mind after research and budgeting, it is better to tell clients this and withdraw from the project. Your honesty will win their respect and may pay off in other jobs down the line. Businesspeople are not stupid. If you buffalo them this time you may sell the project. But you will never sell them another thing as long as you live.

Repeat business and word-of-mouth advertising and recommendations are the lifeblood of the independent producer. Therefore, a few minutes of real soul-searching at this stage of the game can be crucial.

## Making an Effective Presentation—"The Dog and Pony Show"

An effective presentation is a selling tool that can be as important as the bottom line of the budget or the actual proposal. Thus it is imperative that you learn the techniques and know the tips that set a mediocre presentation apart from an excellent one.

The quite respectable, quite necessary and generally final sales pitch made to a client is called "The Dog and Pony Show" by those in the business. When making this final sales pitch, pull out all the stops and use all your powers of showmanship, positive thinking and persuasion to convince the client to buy your deal. So institutionalized has this performance become that we call it the D&P (or D-and-P).

You may be a very good video producer with a bookshelf full of Peabody or Emmy Awards. The idea you are selling may be able to save the client millions of dollars if it is implemented. None of this matters, however, unless you have a solid D&P to sell it all to that client. This presentation is your only chance to convince the client that you are the one person for the job. Following are some elements of a good D&P.

## Setting the Stage

The impression you make at the D&P has to be a lasting one. No matter how casually you may be received by the client, consider this a formal occasion. Dress for it. The client may sit at the table in shirtsleeves, but you must keep your jacket and tie on.

Utilize the salesperson's technique of practicing your speech in front of a mirror. Know all of your facts and figures without having to fumble through sheaves of paper to find them. State how long the presentation will take at the outset and stick to your time frame.

Although this is a formally prepared speech, make it sound as if it's coming off the top of your head. Spontaneity and a casual delivery put people at ease. Make time at the end for questions but try to be so thorough that there are few, if any.

Take charge and maintain it. Although it is the client's time and place, it is your moment. Be polite, cordial, friendly and don't forget to inject a sense of humor into the proceedings.

Remember, too, that you are in the audiovisual business. Make good use of charts, slides, graphs and audio recordings if you can. These allow the client to see your efficiency and your command of the media you represent.

## The Demo Reel

The demonstration, or demo reel, is your major sales tool. It is quite simply a compilation of your best work to date, assembled on one tape. Show it to the prospective client to prove that you know what you're doing.

A smooth and impressive demo reel isn't just thrown together. It requires careful thought and preparation. Here are some suggestions for one that shines:

- *Keep it short*. This is a demo reel, not an epic extravaganza. Keep it no longer than five to ten minutes, which is about the length of time any busy executive can be expected to sit still and pay attention. If you do mainly industrial or training tapes, select brief sections from two or three that show your best creativity. If you do spot TV commercials, choose a few of your best 30s and 60s to showcase here.
- *Orchestrate it*. Treat the demo reel like a complete production. Use musical

segues between sections to create a sense of flow. If portions of the reel need to be explained, use a narrator. Title the beginning and end with tasteful graphics using your company logo. You might even want to run a short credit roll at the end over some triumphal music to show that you aren't working alone. It impresses the client and leaves a feeling of having seen a show rather than a demonstration. It's also about the only time you will be able to use your credits in this business!

• *Tailor it to fit the client.* If you're going to be a generalist, then make more than one demo reel. You may have won awards for your industrial work, but an industrial reel is practically worthless to a fast-food client.

Advertising agencies, especially, tend to pigeonhole producers and directors. If you have a reputation with an agency for doing excellent airline work, for example, you can count on *not* being called to produce its spots for a hotel chain. Therefore, develop a series of demo reels for each of the major areas in which you intend to work. If you just haven't done enough to make separate reels, then show as much diversity as possible on the one reel you put together.

• *Show it off.* Normally you will do the D&P at the client's place of business. You may be making your presentation in an office or a conference room, which means that the kind of equipment you use is critical. It should be easily portable, simple to set up and easy to use.

With recent advances in the quality of 1/2-inch format VCRs, many producers are showing demo reels on VHS or Beta rather than the more cumbersome 3/4-inch. These "home market" consumer machines are very compact and able to stand the abuse of frequent moves. Many come with stereo sound capability for superb audio quality.

The monitor should be the largest screen practicable. Usually more than one person will be viewing your tape, and you want the screen to be commanding. There are a number of large-screen monitors available. Make sure that it is aligned and working properly before the showing. Nothing can kill your demo reel—and your sale—quicker than having to make excuses to the client for a fuzzy, unregistered picture.

If the client comes to your office, make sure that you have a tasteful screening room. A wide variety of very large video projection units is available. By adding high-quality speakers, pretty common video can look and sound very impressive. After all, impressing the client is what the demo reel is all about.

## The Handouts

During the course of the D&P you will hand out copies of the proposal, including the budget. Be sure you know how many people will be at the presentation so you can have copies for everyone. Have the proposal bound as we discussed earlier in this chapter. A nice touch is to have the name of each individual at the presentation printed on the cover. You can do this yourself with press-on letters or with a typed gummed label. This personal touch makes the individual feel important. It also imparts a feeling that you really care about the account.

Give the clients time to read the proposal. You might wish to read through it formally so that each person can follow along. At the conclusion of the proposal reading, open the session to questions and answers. If you've done your job well and presented a package in line with the client's budget limitations, the main question you should anticipate is, "When do we start?"

Give the client and each of the staff members a copy of your brochure with a business card attached. Again, this makes the people at the presentation feel as if you have taken a personal interest in them. Make a brief statement to every person, calling each one by name. For example, "Jack, it's been a pleasure meeting you. Here is a pamphlet for you to keep. Don't hesitate to call if you have any questions. We're looking forward to working with you."

Finally, print up a brief questionnaire (see Figure 1.6) to actively solicit the client's response to the presentation. Explain that you do this to test yourself, not the client. Make it simple for the client to fill out and include a stamped, self-addressed envelope.

Based on your knowledge of the clients, invent a series of questions designed to appeal to them. The questionnaire is obviously a leading one. Its purpose is to make

---

**Figure 1.6: Questionnaire to Solicit Client Response**

1. Did the presentation answer all of your questions about this production? [     ] Yes   [     ] No

2. Did the quality of the demo reel meet your expectations? [     ] Yes   [     ] No

3. On a scale of 1 to 10 please indicate your response to the demo reel.

4. Were all staff members needed to make the decision present? [     ] Yes   [     ] No

5. If not, would you like a second presentation to brief them? [     ] Yes   [     ] No

6. If so, when?_____.

7. We will be making our decision to proceed with production in _____ days.

8. We have the following questions about the production based on issues raised after the presentation:

_____

_____

_____

_____

the clients comfortable and involved so that they lack any reason to turn the project down.

## Closing the Deal

No matter how well you have presented your case, don't expect a decision on the spot. In most cases you are asking clients to commit to an expenditure of several thousand dollars. If this is the client's first venture into video production, the prospect is also a little strange and frightening. They will need some time to think it over; how much time is somewhat up to you.

During your presentation you should plant a firm suggestion that you can't wait forever to get rolling. You have budgeted this project based on several factors, including the current price of equipment, videotape, etc. These figures can change without notice. Your budget is also contingent upon the availability of certain talent, both on and off camera. Explain these factors to the client. Finally, you can schedule only so much of your time for this one project; there are others waiting in the wings. Businesspeople will certainly understand these considerations.

There is a phenomenon known as "cooling off" which takes place in potential buyers. The longer they have to think over a prospective major purchase, the more likely they are to pass on it. In the heat of a good sales pitch, people may sign a contract to buy an expensive car or a resort condominium, even if they really can't afford it. That's why many states have passed laws mandating a cooling-off period, giving clients the right to cancel such contracts for a period of time following the sale.

This cooling can take place in clients if you give them more than a few days to decide. Five working days are generally enough for anyone to make a yes-or-no decision. During those five days you can perform follow-up maneuvers to help the client make a favorable decision.

### Follow-up Actions

First, send a letter to arrive the day after the presentation. This is a simple exercise in good manners. Thank the client for the time and opportunity to make the presentation. Tell them how enjoyable it was for you to meet the staff and with what pleasure you are looking forward to working with them on the project. Write this letter as if the project is definitely going to proceed. Keep it positive and cordial. Use phrases like, "When we begin production, etc." instead of, "If we begin production, etc." Always plant *positive* suggestions. These help keep the client pointed toward the actuality of going ahead.

Next, evaluate your performance at the D&P. Some producers tape the entire presentation on a small, concealable audio recorder. Playing it back gives them insight into their own strengths and weaknesses. It also lets them hear the audience's reactions for a second time, which can be helpful in preparing for questions that may arise.

Two days after the D&P, make a telephone call. Using the information from your playback, you might impress the client by referring back to one or two areas of concern expressed at the meeting. If the people you speak to have not decided by this point, inform them that you will phone back on Friday (at a specific time) to receive the go-ahead.

If the answer does not come on Friday, press firmly for reasons why. If they need the weekend to decide, that is fine, but on Monday you must have the final word. If you don't receive it then, forget about this client and move on. You have encountered one of the frustrations of this business. You have done your best, but the client won't move.

## Negotiations

Sometimes you will encounter a client who is a horse trader. This person has an innate need to haggle over the cost of everything. No matter how carefully you have constructed a case for your budget, this client isn't going to go for it unless you make some concessions.

Other clients will have absolutely fixed financial limitations that prevent them from buying your product at the price you've quoted, even though they may want the production very much. With both of these types, you must be prepared to negotiate.

The very word "negotiate" means to compromise. You give up somthing; the client gives up something. You reach an equitable agreement and the production proceeds.

Anticipating the need to negotiate on price, many producers build in an additional "fudge factor" to the budget. This percentage varies from 2% to as much as 10%. This practice bothers many people from an ethical standpoint. If the client buys the budget you have proposed from the outset in one of these cases, you've made an additional windfall profit. While this may make you happy, it will backfire if word gets out to the industry that you are a rip-off artist.

A better tactic—one that will help maintain your good reputation in a fairly small, close-knit business—is to prepare an alternative budget. Demonstrate to the client that you can cut costs, but in so doing, you will also be eliminating certain positive production values.

For example, you might handle the haggler this way: "Mr. Jones, I understand that you want a high-quality piece here, and our budget has provided for that quality. I really can't cut back on the essentials, but we could make the piece shorter if you like. Every day we cut out of the schedule saves you X dollars, as you can see. We might also be able to reduce the cost somewhat by eliminating the music or changing it from an original composition to something from a stock music library."

Be able to make your case in specific terms. Point out exactly what the sacrifice in dollars will mean to the finished product. A negotiation has to be an open and honest exchange of viewpoints on both sides. Each party must be willing to give and take. Only as a last resort should you consider reducing your fair profit margin on the project. If landing the account might mean a great deal more work down the line from this client, it may be wise to reduce your profit on this one project to get the contract. Make clear, however, your reason for backing down and let the client know in no uncertain terms that this is a one-time deal.

## Expectations

Selling, which is really what we've been discussing here, is a numbers game. The more sales calls you make, the higher the volume of business you can expect to do.

No matter how well you prepare and present a D&P, don't expect it to result in an automatic sale every time. Avoid blaming yourself for failed prospects. Even the people at the very top cannot and do not expect to sell every idea they have.

The biggest frustration you will encounter as an independent producer will be knowing in your heart that the project you've proposed to clients is good for them and the best thing available, and that the price is rock bottom. Yet they won't buy it. There is just no accounting in the end for what motivates people. That's why selling is a fine art, not a science.

Learn to expect and accept rejection and do not let it turn you into a cynic. The times the sale does go through—when the client shakes your hand and signs the contract—are among the peak experiences for the independent producer. And in the long run, they will more than make up for the failures.

## SUMMARY

In this chapter we have presented one method for the producer to use in achieving one of the most important functions in independent business: finding a client and making a sale. Knowing how to sell your idea, as well as the specifics of your proposal, is an extreme asset for the independent producer. Only when you become well established will clients come to you; in the meantime, you must seek them out and know how to sell your product.

## NOTES

1. Comments from Saul Turtletaub and the other Hollywood producers quoted in the remainder of this book are taken from a series of taped interviews conducted by the author each January from 1980 to 1985 as part of an ongoing research program for the Radio-TV-Film major at the University of Wisconsin-Oshkosh. This quote is taken from an interview conducted in January, 1984. Further information on this project is described in articles by the author in the Summer/Fall 1981 and the Winter/Spring 1984-85 editions of *American Cinemeditor* mag-

azine—a publication of the Honorary Professional Society, American Cinema Editors, Inc. Copies may be obtained by writing American Cinema Editors, Inc., 4416½ Finley Ave., Los Angeles, CA 90027.

2. You can find a list of most of the major trade associations in America in *The Encyclopedia of Associations* (Detroit, MI: Gale Research Co., 1985).

3. For a detailed discussion of budgeting procedures, see Richard E. Van Deusen, *Practical AV/Video Budgeting* (White Plains, NY: Knowledge Industry Publications, Inc., 1984).

4. Gene Roddenberry, interview with author, January, 1982.

5. For a complete list of video equipment dealers, see *The Video Register 1985-86,* by the editors at Knowledge Industry Publications, Inc. (White Plains, NY: Knowledge Industry Publications, Inc., 1985).

6. For help in this area of post-production, see Gary H. Anderson, *Video Editing and Post-Production* (White Plains, NY: Knowledge Industry Publications, Inc., 1984).

7. For a list of music and sound-effects libraries, and information concerning contacts for celebrities, services, networks, etc., see *The Hollywood Reporter Studio Blu-Book Directory, 1985* (Hollywood, CA: Verdugo Press, 1985).

# Appendix to Chapter 1

---

## PICTURE BUDGET DETAIL

TITLE    NASHVILLE WITH A BULLET      PICTURE-NO. 01-80-1

GLENARM PRODUCTIONS      DATE PREPARED October 1,    . 19 80

| ACCOUNT NUMBER | DESCRIPTION | | TOTAL | TOTALS | TOTALS |
|---|---|---|---|---|---|
| 1 | Story | | | | 150,000 |
| 2 | Continuity and Treatment | | | | 51,500 |
| 3 | Producer | | | | 515,000 |
| 4 | Director | | | | 180,000 |
| 5 | Cast | | | | 819,500 |
| 6 | Bits | | | | 77,700 |
| 7 | Extras | | | | 72,300 |
| | | | | | |
| | | | | | |
| | Sub Total | | | | 1,866,000 |
| 8 | Production Staff Salaries | | | | 92,000 |
| 9 | Production Operating Staff | | | | 380,800 |
| 10 | Set Designing | | | | 92,200 |
| 11 | Set Operation Expenses | | | | 222,012 |
| 12 | Cutting - Film - Laboratory | | | | 279,999 |
| 13 | Music | | | | |
| 14 | Sound | | | | |
| 15 | Transportation - Studio | | | | 46,000 |
| 16 | Location | | | | 308,643 |
| 17 | Studio Rental | | | | 36,400 |
| 18 | Tests and Retakes | | | | 23,000 |
| 19 | Publicity | | | | 90,000 |
| 20 | Miscellaneous | | | | 1,500 |
| 21 | Insurance - Taxes - Licenses and Fees | | | | 417,850 |
| 22 | General Overhead | | | | 80,900 |
| | | | | | |
| | Sub Total | | | | 2,071,304 |
| | | | | SUB GRAND TOTAL | 3,937,304 |
| | Grand Total | Contingency factor: | | 590,595 = | 4,527,899 |

Approved   Gary Lynn Hall                Producer

Prepared From 156      Page Script Dated July, 1980

   60     Day Shooting Scheduled at Nashville Location & L.A.     Studio

Director Paul Krasny

Budget by

Robert M. Jacobs, Ph.D.

ENTERPRISE STATIONERS 7401 SUNSET L.A. CA. 90046 (213) 876-3533

Picture Budget Detail forms reproduced with the permission of Enterprise Stationers.

Page No.

TITLE  NASHVILLE WITH A BULLET         PICTURE NO.

GLENARM PRODUCTIONS
                              DATE PREPARED  May 1, 1980

| ACCOUNT NUMBER | DESCRIPTION | DAYS, WKS, OR QUANTITY | RATE | TOTALS |
|---|---|---|---|---|
| 1 | STORY - | | | |
| | A.  STORY PURCHASE | | | 150,000 |
| | B.  TITLE PURCHASE | | | |
| | | | | |
| | TOTAL STORY | | | 150,000 |
| 2 | CONTINUITY AND TREATMENT | | | |
| | A.  WRITERS | | | 50,000 |
| | B.  STENOGRAPHER | | | |
| | C.  MIMEOGRAPH EXPENSE | | | 500 |
| | D.  RESEARCH EXPENSE | | | 1,000 |
| | | | | |
| | TOTAL CONTINUITY AND TREATMENT | | | 51,500 |
| 3 | PRODUCER | | | |
| | A.  PRODUCER | | | 150,000 |
| | B.  ASST. PRODUCER | | | 50,000 |
| | C.  SECRETARIES | | | 15,000 |
| | d.  Executive Producers | | | 300,000 |
| | | | | |
| | TOTAL PRODUCER | | | 515,000 |
| 4 | DIRECTOR | | | |
| | A.  DIRECTOR | | | 150,000 |
| | B.  SECRETARIES | | | 15,000 |
| | C.  PENSION CONTRIBUTIONS | | | 15,000 |
| | | | | |
| | TOTAL DIRECTORS | | | 180,000 |

ENTERPRISE STATIONERS. HOLLYWOOD

Page 8

| TITLE | NASHVILLE WITH A BULLET | | PICTURE NO. | | |
|---|---|---|---|---|---|

DATE PREPARED   May 1, 1980

| ACCOUNT NUMBER | DESCRIPTION | DAYS, WKS, OR QUANTITY | RATE | | TOTALS |
|---|---|---|---|---|---|
| 5 | CAST | | | | |
| | Male lead   John | 12 weeks | | | 150,000 |
| | Female lead   Lynn | " | | | 150,000 |
| | 2nd Male lead Pappy | | | | 100,000 |
| | Female supporting Dixie | | | | 40,000 |
| | Male supporting   B.J. | | | | 100,000 |
| | Male supporting   Tony | | | | 60,000 |
| | Angelo | | | | 40,000 |
| | James | | | | 40,000 |
| | Paul | | | | 25,000 |
| | Genova | | | | 25,000 |
| | Cox | | | | 5,000 |
| | T.G. | | | | 5,000 |
| | Doyle | | | | 5,000 |
| | BUYOUTS | | 125% | | |
| | PENSION H&W CONTRIBUTIONS | | | | 74,500 |
| | TOTAL CAST | | | | 819,500 |
| 6 | BITS | | | | |
| | Narrator | | | | |
| | Mifflin | | | | 5,000 |
| | Bartender | | | | 2,500 |
| | Rob | | | | 10,000 |
| | Del | | | | 5,000 |
| | Poolplayer | | | | 1,000 |
| | Girl Singer | | | | 2,500 |
| | Guitar picker | | | | 2,500 |
| | Dean Duncan | | | | 10,000 |
| | Old Man at Party | | | | 5,000 |
| | Sgt. Baker | | | | 2,500 |
| | Tina | | | | 1,500 |
| | Doc | | | | 5,000 |
| | BUYOUTS | | 125% | | |
| | PENSION CONTRIBUTIONS | | | | 5,2 |
| | OVERTIME ON BITS | | | | 10,000 |
| | FITTING CHARGES | | | | 10,000 |
| | TOTAL BITS | | | | 77,700 |

Page No. _____

TITLE    NASHVILLE WITH A BULLET                    PICTURE NO. _____

                                          DATE PREPARED    May 1, 1980

| ACCOUNT NUMBER | DESCRIPTION | DAYS, WKS, OR QUANTITY | RATE | | TOTALS |
|---|---|---|---|---|---|
| 7 | EXTRAS | 150. | 00 | | |
| | Party- 20 | | | | 6,000 |
| | Jamboree- 50 | | | | 15,000 |
| | Street Scenes-10 | | | | 4,500 |
| | Bar-5 | | | | 1,500 |
| | Studio-5 | | | | 1,500 |
| | Murder Scene #1-10 | | | | 4,500 |
| | Murder Scene #2-6 | | | | 1,800 |
| | Murder Scene #3-5 | | | | 1,500 |
| | Murder Scene #4-5 | | | | 1,500 |
| | Police Station-10 | | | | 4,500 |
| | | | | | |
| | OVERTIME FOR EXTRAS | | | | 15,000 |
| | FITTING FOR EXTRAS | | | | 5,000 |
| | SERVICE FEES FOR EXTRAS | | | | 10,000 |
| | ADJUSTMENTS FOR EXTRAS . | | | | |
| | STAND INS | | | | |
| | SCHOOL TEACHER'S | | | | |
| | STUNT PEOPLE (see special effects) | | | | |
| | STUNT ADJUSTMENTS | | | | |
| | TOTAL EXTRAS | | | | 72,300 |

Page No. _____

TITLE    NASHVILLE WITH A BULLET      PICTURE NO. _____

DATE PREPARED    May 1, 1980

| ACCOUNT NUMBER | DESCRIPTION | DAYS, WKS, OR QUANTITY | RATE | TOTAL |
|---|---|---|---|---|
| 8 | PRODUCTION STAFF SALARIES | | | |
| | A.   PRODUCTION MANAGER | 12 week | | 24,000 |
| | B.   UNIT MANAGER | | | 12,000 |
| | C.   1ST ASST. DIRECTOR | | | 12,000 |
| |      SEVERANCE | | | |
| | D.   2ND ASST. DIRECTOR | | | 6,000 |
| |      SEVERANCE | | | |
| | E.   EXTRA ASST. DIRECTORS | | | |
| | F.   SECRETARIES | | | 6,000 |
| | G.   DIALOGUE CLERK | | | |
| | H.   SCRIPT CLERK | | | 8,000 |
| | U.   DANCE DIRECTOR | | | |
| | J.   CASTING DIRECTOR & STAFF | | | |
| | K.   TECHNICAL ADVISOR | | | |
| | L.   FIRST AID | | | 12,0 |
| | M.   LOCATION AUDITOR | | | 12,000 |
| | **TOTAL PRODUCTION STAFF** | | | 92,000 |
| 9 | PRODUCTION OPERATING STAFF | | | |
| | A.   CAMERAMEN | | | |
| |    1.   1ST CAMERAMAN   D.P. | 12 week | 1200 | 14,400 |
| |    2.   CAMERA OPERATORS (2) @ 900 each | 12 | 900 | 21,600 |
| |    3.   FOCUS ASST. CAMERAMEN (3) @ 450 each | | | |
| |    4.   ASST. CAMERAMEN | 12 week | 450 | 5,400 |
| |    5.   CAMERA MECHANICS | | | |
| |    5.   COLOR DIRECTOR | | | |
| |    7.   STILL MAN | 12 week | 300 | 3,600 |
| |    8.   STILL GAFFER | | | |
| |    9.   PROCESS CAMERAMAN | | | |
| |   10.   ASST. PROCESS CAMERAMAN | | | |
| |   11.   EXTRA CAMERA OPERATORS | | | |
| |   12.   EXTRA CAMERA ASSISTANTS | | | |
| |   13.   O.T. CAMERA CREW 43-48 HOURS | | | 10,000 |
| | **TOTAL ACCT. 9-A** | | | 55,000 |

ENTERPRISE STATIONERS

Page No. 5

TITLE   NASHVILLE WITH A BULLET

PICTURE NO.

DATE PREPARED   May 1, 1980

| ACCOUNT NUMBER | DESCRIPTION | DAYS, WKS. OR QUANTITY | RATE | TOTALS |
|---|---|---|---|---|
| 9 | PRODUCTION OPERATING STAFF (Contd.) | | | |
| | B.   SOUND DEPT. | | | |
| | 1.   MIXER | 12 week | 1500 | 18,000 |
| | 2.   RECORDER | | | |
| | 3.   BOOM MAN | " | 800 | 9,600 |
| | 4.   CABLEMAN | | | |
| | 5.   CABLE BOOM MAN | | | |
| | 6.   P.A. SYSTEM OPERATOR | | | |
| | 7.   DREAM OPERATOR | | | |
| | 8.   SOUND MAINTENANCE | | | |
| | TOTAL ACCT. 9-B | | | 27,600 |
| | C.   WARDROBE DEPT. | | | |
| | 1.   WARDROBE DESIGNER | 6 week | 500 | 3,000 |
| | 2.   WARDROBE BUYER | | | |
| | 3.   1ST WARDROBE GIRL | 12 | 500 | 6,000 |
| | 4.   2ND WARDROBE GIRL | 12 | 300 | 3,600 |
| | 5.   1ST WARDROBE MAN | | | |
| | 6.   2ND WARDROBE MAN | | | |
| | 7.   TAILOR | | | |
| | 8.   SEAMSTRESS | 12 | 300 | 3,600 |
| | 9.   EXTRA HELP | | | |
| | TOTAL ACCT. 9-C | | | 16,200 |
| | D.   MAKE-UP AND HAIRDRESSING | | | |
| | 1.   HEAD MAKE-UP MAN | 12 wks | 800 | 9,600 |
| | 2.   2ND MAKE-UP MAN | 12 wks | 500 | 6,000 |
| | 3.   HEAD HAIRDRESSER | 12 wks | 600 | 7,200 |
| | 4.   2ND HAIRDRESSER | 12 wks | 400 | 4,800 |
| | 5.   BODY MAKE-UP GIRL | | | |
| | 6.   EXTRA HELP | | | |
| | TOTAL ACCT. 9-D | | | 27,600 |

Page No. ___

| | | | |
|---|---|---|---|
| TITLE | NASHVILLE WITH A BULLET | PICTURE NO. | |
| | | DATE PREPARED | May 1, 1980 |

| ACCOUNT NUMBER | DESCRIPTION | DAYS, WKS, OR QUANTITY | RATE | TOTALS |
|---|---|---|---|---|
| | **PRODUCTION OPERATING STAFF (Contd.)** | | | |
| | E.   GRIP DEPT. | | | |
| | 1.   1st GRIP | 12 wks | 1000 | 12,000 |
| | 2.   BEST BOY | 12 | 800 | 9,600 |
| | 3.   SET OPERATION GRIPS   6 each | 12 | 500 | 36,000 |
| | 4.   EXTRA LABOR | | | |
| | 5.   CAMERA BOOM OPERATORS | | | |
| | 6.   CRAB DOLLY GRIP | 12 | 500 | 6,000 |
| | TOTAL ACCT. 9-E | | | 63,600 |
| | F.   PROPERTY | | | |
| | 1.   HEAD POOPERTY MAN | 14 | 800 | 11,200 |
| | 2.   2nd PROPERTY MAN | 12 | 500 | 6,000 |
| | 3.   3rd PROPERTY MAN | 12 | 300 | 3,600 |
| | 4.   OUTSIDE HELP | | | |
| | 5.   EXTRA HELP | | | |
| | TOTAL ACCT. 9-F | | | 20,800 |
| | G.   SET DRESSING DEPT. | | | |
| | 1.   HEAD SET DRESSER | 12 | 800 | 9,600 |
| | 2.   ASST. SET DRESSER | 12 | 500 | 6,000 |
| | 3.   SWING GANG | | | |
| | 4.   DRAPERY MAN | | | |
| | 5.   ASST. DRAPERY MAN | | | |
| | 6.   NURSERY MAN | | | |
| | 7.   EXTRA LABOR | | | |
| | TOTAL ACCT. 9-G | | | 15,600 |
| | H.   ELECTRICAL DEPT. | | | |
| | 1.   GAFFER | 12 | 1200 | 14,400 |
| | 2.   BEST BOY | 12 | 800 | 9,600 |
| | 3.   ELECTRICAL OPERATING LABOR   2 each | 12 | 500 | 12,000 |
| | 4.   GENERATOR OPERATOR | | | |
| | 5.   ELECTRICAL MAINTENANCE MAN | | | |
| | 6.   RIGGING & STRIKING CREW   2 each | 12 | 500 | 12,00 |
| | 7.   WIND MACHINE OPERATOR | | | |
| | TOTAL ACCT. 9-H | | | 48,000 |

Page No. 7

TITLE  NASHVILLE WITH A BULLET

PICTURE NO.

DATE PREPARED  May 1, 1980

| ACCOUNT NUMBER | DESCRIPTION | DAYS, WKS, OR QUANTITY | RATE | TOTALS |
|---|---|---|---|---|
| 9 | PRODUCTION OPERATING STAFF (Contd.) | | | |
| | I.  LABOR DEPT. | | | |
| | 1.  STANDBY LABORER | | | |
| | 2.  ASST. LABORERS | | | |
| | TOTAL ACCT. 9-I | | | |
| | J.  SPECIAL EFFECTS | | | |
| | 1.  HEAD SPECIAL EFFECTS MAN | 4 week | 1000 | 4,000 |
| | 2.  ASST. SPECIAL EFFECTS MAN | 4 | 600 | 2,400 |
| | 3.  PLUMBER | | | |
| | TOTAL ACCT. 9-J | | | 6,400 |
| | K.  SET STANDBY OPERATORS | | | |
| | 1.  CARPENTER | 4 | 1000 | 4,000 |
| | TOTAL ACCT. 9-K | | | 4,000 |
| | L.  SET STANDBY PAINTERS | | | |
| | 1.  PAINTER | 4 | 1000 | 4,000 |
| | TOTAL ACCT. 9-L | | | 4,000 |
| | M.  SET WATCHMAN | | | |
| | 1.  WATCHMEN | | | |
| | TOTAL ACCT. 9-M | | | |
| | N.  WRANGLERS | | | |
| | 1.  S.P.C.A. MAN | | | |
| | 2.  HEAD WRANGLER | | | |
| | 3.  WRANGLERS | | | |
| | O.  MISCELLANEOUS | | | |
| | GRAND TOTAL SET OPERATING SALARIES | | | 380,800 00 |

TITLE NASHVILLE WITH A BULLET                PICTURE NO. _____          Page No. _

DATE PREPARED _____

| ACCOUNT NUMBER | DESCRIPTION | TIME | RATE | | | TOTAL |
|---|---|---|---|---|---|---|
| 10 | SET CONSTRUCTION | | | | | |
| | A. Art Director | 6 weeks | 800 | | | 4,800 |
| | B. Asst. Art Director | 6 | 300 | | | 1,800 |
| | C. Sketch Artist | | 300 | | | 1,800 |
| | D. Draftsman | | 300 | | | 1,800 |
| | E. Set Supervisor | | | | | |
| | F. Material & Supplies | | | | | 20,000 |
| | G. Construction Supervisor | | 600 | | | 3,600 |
| | H. Miscellaneous | | | | | 10,000 |
| | | LABOR | MATERIAL | | | |
| 1 | Rigging & Striking | | | | | 8,000 |
| 2 | Trucking & Hauling | | | | | 8,000 |
| 3 | Painter | | | | | 12,000 |
| 4 | | | | | | |
| 5 | | | | | | |
| 6 | | | | | | |
| 7 | | | | | | |
| 8 | | | | | | |
| 9 | | | | | | |
| 10 | | | | | | |
| 11 | | | | | | |
| 12 | | | | | | |
| 13 | | | | | | |
| 14 | | | | | | |
| 15 | | | | | | |
| 16 | | | | | | |
| 17 | | | | | | |
| 18 | | | | | | |
| 19 | | | | | | |
| 20 | | | | | | |
| 21 | | | | | | |
| 22 | | | | | | |
| 23 | | | | | | |
| 24 | | | | | | |
| 25 | | | | | | |
| 26 | | | | | | |
| 27 | | | | | | |
| 28 | | | | | | |
| 29 | | | | | | |
| 30 | | | | | | |
| 31 | | | | | | |
| 32 | | | | | | |
| 33 | | | | | | |
| 34 | | | | | | |
| 35 | | | | | | |
| 36 | | | | | | |
| 37 | | | | | | |
| 38 | | | | | | |
| 39 | | | | | | |
| | Rigging Labor Grip | | 400 | | | 2,400 |
| | Striking | | | | | 5,000 |
| | Backings | | | | | 8,000 |
| | Greens | | | | | 5,000 |
| | TOTAL SETS | | | | | 92,200 |

ENTERPRISE STATIONERS

Page No. _____

TITLE   NASHVILLE WITH A BULLET _____   PICTURE NO. _____

DATE PREPARED   May 1, 1980 _____

| ACCOUNT NUMBER | DESCRIPTION | DAYS, WKS, OR QUANTITY | RATE | TOTALS |
|---|---|---|---|---|
| 11 | SET OPERATION EXPENSES | 12 week | | |
| | A.   Camera Equipment Rentals | | | 16,380 |
| | B.   Camera Equipment Purchases | | | 4,000 |
| | C.   Camera Car Rentals | | | 3,300 |
| | D.   Camera Crane Rentals | | | 4,632 |
| | E.   Wardrobe Purchased | | | 10,000 |
| | F.   Wardrobe Rentals | | | 5,000 |
| | G.   Wardrobe Maintenance | | | 5,000 |
| | H.   Grip Equipment Rented | | | 18,400 |
| | I.   Prop Equipment Rented | | | |
| | J.   Props Purchased | | | 15,000 |
| | JJ.  Prop Man's Petty Cash Exp. | | | 5,000 |
| | K.   Props Rented & returned | | | 8,000 |
| | L.   Props - Loss & Damaged | | | 8,000 |
| | M.   Set Dressing Rentals | | | 12,000 |
| | N.   Set Dressing Purchased | | | 12,000 |
| | O.   Draperies Purchased & Rented | | | 4,000 |
| | P.   Nursery - Purchased & Rented | | | 2,500 |
| | Q.   Process Equipment Rentals | | | |
| | R.   Make-up Purchases | | | 5,000 |
| | S.   Hairdressing Purchases & Rentals | | | 5,000 |
| | T.   Electrical Equipment Rentals | | | 34,200 |
| | U.   Electrical Equipment Purchased | | | 4,000 |
| | V.   Electrical Power | | | 8,000 |
| | W.   Rentals on Picture Cars - Trucks - Planes - Wagons - Livestock, etc. | | | 10,000 |
| | X.   Miscellaneous Rentals & Purchases | | | 8,000 |
| | Y.   Generator Rental - Gas & Oil - | | | 4,000 |
| | Z.   Special Effect Purchases & Rentals | | | 10,000 |
| | Total Set Operation Expense | | | 222,012 |

Page No.

| ACCOUNT NUMBER | DESCRIPTION | DAYS, WKS, OR QUANTITY | RATE | TOT |
|---|---|---|---|---|

TITLE NASHVILLE WITH A BULLET          PICTURE NO.

DATE PREPARED

| ACCOUNT NUMBER | DESCRIPTION | DAYS, WKS, OR QUANTITY | RATE | TOT |
|---|---|---|---|---|
| 12 | CUTTING FILM LABORATORY | | | |
| | Supervising Editor | | | |
| | A. EDITOR | 16 | 500 | 24,000 |
| | B. ASST. CUTTER | | 500 | 8,000 |
| | C. SOUND CUTTER | 8 | 700 | 5,600 |
| | D. MUSIC CUTTER | 4 | 700 | 2,800 |
| | E. NEGATIVE CUTTER ( 10 reels ) | 10 | 225 | 2,250 |
| | F. NEGATIVE ACTION RAW STOCK  100,000 feet | | | 28,500 |
| | G. NEGATIVE SOUND RAW STOCK | | | 12,000 |
| | GG. TAPE RENTAL | | | |
| | H. DEVELOP ACTION | | | 24,000 |
| | HH. DEVELOP SOUND ( mag transfer $ stock) | | | 7,000 |
| | I. PRINT ACTION | | | 20,000 |
| | II. PRINT SOUND | | | 12,0 |
| | J. MAGNASTRIPE - PRODUCTION | | | |
| | JJ. MAGNASTRIPE - SCORE & DUBBING | | | |
| | K. COLOR SCENE PILOT STRIPS | | | |
| | KK. 16MM COLOR PRINTS (FROM CCO) | | | |
| | KKK. INTER-NEGATIVE | | | 18,000 |
| | L. SEPARATION MASTERS | | | |
| | LL. INTER-POSITIVE  CRI | | | 30,000 |
| | M. ANSWER PRINT | | | 12,000 |
| | MM. COMPOSITE PRINT | | | 4,000 |
| | N. FINE GRAIN PRINT | | | |
| | NN. PANCHROMATIC (FG) | | | |
| | NNN. 16MM PRINTS | | | |
| | O. FADES-DISSOLVES-DUPES & FINE GRAIN | | | 20,000 |
| | OO. REPRINTS | | | |
| | P. TITLES, MAIN & END | | | 14,000 |
| | Q. CUTTING ROOM RENTAL | | | 6,0 |
| | R. CODING | | | 2,200 |

ENTERPRISE STATIONERS, HOLLYWOOD 5-2540
NO. 88

Page No. 11

TITLE   NASHVILLE WITH A BULLET

PICTURE NO.

DATE PREPARED   May 1, 1980

| ACCOUNT NUMBER | DESCRIPTION | DAYS, WKS, OR QUANTITY | RATE | TOTALS |
|---|---|---|---|---|
| 12 | CUTTING FILM LABORATORY (Contd.) | | | |
| | R.   PROJECTION | | | 3,000 |
| | S.   MOVIOLA RENTALS | 8 week | | 3,000 |
| | T.   REELS & LEADER | | | 800 |
| | U.   CUTTING ROOM SUPPLIES   Rental | | | 2,500 |
| | Purchase | | | 2,500 |
| | V.   STOCK SHOTS | | | |
| | W.   PROCESS PLATES | | | |
| | X.   SALES TAX | | | 15,849 |
| | XX.  CODING | | | |
| | LABORATORY SUB-TOTAL | | | 279,999 |
| | TOTAL CUTTING FILM LABORATORY | | | SAME |

ENTERPRISE STATIONERS, HOLLYWOOD
   NO. 95

| TITLE | NASHVILLE WITH A BULLET | PICTURE NO. | | |
|---|---|---|---|---|
| | | DATE PREPARED | May 1, 1980 | |

| ACCOUNT NUMBER | DESCRIPTION | DAYS, WKS, OR QUANTITY | RATE | TOTALS |
|---|---|---|---|---|
| 13 | MUSIC | | | |
| | A.  Music Supervisor | | | |
| | B.  Director | | | |
| | C.  Composer/conductor/arranger | | | |
| | D.  Musicians  20 for 10 days @ 130 each | | | |
| | E.  Singers | | | |
| | F.  Arrangers | | | |
| | G.  Copyists | | | |
| | H.  Royalties | | | |
| | I.  Purchases | | | |
| | J.  Miscellaneous | | | |
| | K.  Instrument Rental & Cartage | | | |
| | L.  Librarian | | | |
| | Total Music | | | |
| 14 | SOUND | | | |
| | A.  Royalties | | | |
| | B.  Dubbing Room Rental | | | |
| | C.  Pre-Score Equipment Rentals | | | |
| | D.  Scoring Equipment Rentals | | | |
| | E.  Labor for Dubbing & Etc. | | | |
| | F.  Sound Equipment Rentals | | | |
| | G.  Miscellaneous | | | |
| | H.  Transfer Time | | | |
| | I  Recording tape and batteries | | | |
| | J. Fisher Sound Room | | | |
| | Total Sound | | | |
| 15 | TRANSPORTATION STUDIO | | | |
| | A.  Labor | | | 8,000 |
| | B.  Car Rentals | | | 12,000 |
| | C.  Truck Rentals | | | 4,000 |
| | D.  Bus Rentals | | | 4,000 |
| | E.  Car Allowance | | | 10,000 |
| | F.  Miscellaneous | | | 4,000 |
| | G.  Gas & Oil, - Generator - Mileage | | | 4,000 |
| | H.  Wranglers Cars | | | |
| | I.  Livestock Transportation | | | |
| | Total Transportation | | | 46,000 |

Page No. 1

TITLE    NASHVILLE WITH A BULLET

PICTURE NO.

DATE PREPARED    May 1, 1980

| ACCOUNT NUMBER | DESCRIPTION | DAYS, WKS. OR QUANTITY | RATE | TOTALS | |
|---|---|---|---|---|---|
| 16 | LOCATION | | | | |
| | A.  TRAVELING  airfare | | | 9,000 | |
| | B.  HOTEL | | | 42,000 | |
| | C.  MEALS | | | 60,000 | |
| | D.  LOCATION SITES RENTAL | | | 25,000 | |
| | E.  SPECIAL EQUIPMENT | | | 10,000 | |
| | F.  CAR RENTALS    4 | | | 4,500 | |
| | G.  BUS RENTALS | | | 6,800 | |
| | H.  TRUCK RENTALS (prop, Honeywagon, mis) | | | 20,000 | |
| | I.  SUNDRY EMPLOYEES (drivers, firemen) | | | 12,000 | |
| | J.  LOCATION OFFICE RENTAL | | | 2,200 | |
| | K.  GRATUITIES | | | 5,000 | |
| | L.  MISCELLANEOUS (medical-ambulance) | | | 10,000 | |
| | M.  SCOUTING & PRE-PRODUCTION | | | 51,143 | |
| | N.  POLICE SERVICES & PERMITS | | | 10,000 | |
| | O.  CONTACT MAN (location finder) | | | 4,000 | |
| | TOTAL LOCATION | | | 308,643 | |
| 17 | STUDIO RENTALS | | | | |
| | A.  STAGE SPACE | | | 8,000 | |
| | B.  STREET RENTALS | | | 10,000 | |
| | C.  TEST | | | 2,500 | |
| | D.  VACATION ALLOWANCE (STUDIO) | | | 2,500 | |
| | E.  SURCHARGE ON RENTALS & STUDIO CHARGES | | | 2,200 | |
| | F.  MISCELLANEOUS EXPENSES | | | 4,000 | |
| | G.  DRESSING ROOMS - PORTABLE | | | 5,000 | |
| | H.  OFFICE RENTALS | | | 2,200 | |
| | TOTAL STUDIO RENTALS | | | 36,400 | |
| 18 | TESTS & RETAKES | | | | |
| | A.  TESTS PRIOR TO PRODUCTION | | | 1,000 | |
| | B.  TESTS DURING PRODUCTION | | | 2,000 | |
| | C.  RETAKES AFTER PRINCIPAL PHOTOGRAPHY | | | 8,000 | |
| | D.  PRE-PRODUCTION EXPENSE OR SHOOTING | | | 12,000 | |
| | TOTAL TESTS & RETAKES | | | 23,000 | |

ENTERPRISE STATIONERS, HOLLYWOOD

Page No. 1

TITLE   NASHVILLE WITH A BULLET

PICTURE NO.

DATE PREPARED   May 1, 1980

| ACCOUNT NUMBER | DESCRIPTION | DAYS, WKS. OR QUANTITY | RATE | | TOTALS | |
|---|---|---|---|---|---|---|
| 19 | PUBLICITY | | | | | |
| | A.  ADVERTISING | | | | 50,000 | |
| | B.  UNIT PUBLICITY MAN | | | | | |
| | C.  ENTERTAINMENT | | | | 15,000 | |
| | D.  TRADE AND NEWSPAPER SUBSCRIPTIONS | | | | | |
| | E.  PUBLICITY STILLS SALARIES | | | | | |
| | F.  PUBLICITY STILLS SUPPLIES EQUIPMENT | | | | 2,500 | |
| | G.  PUBLICITY STILLS LAB. CHARGES | | | | 5,000 | |
| | H.  STILL GALLERY RENTAL & EXPENSE | | | | | |
| | I.  Trailer | | | | 8,000 | |
| | J.  PRESS PREVIEW EXPENSE | | | | 4,000 | |
| | K.  SUPPLIES, POSTAGE AND EXPRESS | | | | 3,500. | |
| | L.  MISCELLANEOUS | | | | 2,000 | |
| | TOTAL PUBLICITY | | | | 90,000 | |
| 20 | MISCELLANEOUS | | | | | |
| | A.  VACATION ALLOWANCE | | | | | |
| | B.  RETROACTIVE WAGE CONTINGENCY | | | | | |
| | C.  SUNDRY UNCLASSIFIED EXPENSE | | | | | |
| | D.  COSTS IN SUSPENSE | | | | | |
| | E.  SET COFFEE | | | | 1,000 | |
| | F.  WATER & ICE | | | | 500 | |
| | TOTAL MISCELLANEOUS | | | | 1,500 | |
| 21 | INSURANCE, TAXES, LICENSE AND FEES | | | | | |
| | A.  CAST INSURANCE | | | | | |
| | B.  NEGATIVE INSURANCE | | | | | |
| | C.  LIFE INSURANCE | | | | | |
| | D.  MISCELLANEOUS INSURANCE Total Inclusive | | | | 90,000 | |
| | E.  COMPENSATION & PUBLIC LIABILITY INS. | | | % | | |
| | F.  SOCIAL SECURITY TAX | | 8.5 % | 118,579 | 2 |
| | G.  PERSONAL PROPERTY TAX | | | | | |
| | H.  MISCL. TAXES AND LICENSES | | | | 8,000 | |
| | I.  CODE CERTIFICATE - MPPA | | | | 5,000 | |
| | J.  CITY TAX AND LICENSE | | | | | |
| | K.  UNEMPLOYMENT TAX | | 6 % | 83,703 | |
| | L.  PENSION PLAN CONTRIBUTION ACTORS DIRECTORS WRITERS | | 4.5 | 46,791 | |
| | M.  HEALTH & WELFARE CONTRIBUTION | | | | | |
| | N.  PENSION PLAN - CRAFTS | | 4.5 | 62,7 | 2 |
| | TOTAL A/C 21 | | | | 417,850 | 5 |

ENTERPRISE STATIONERS, HOLLYWOOD

Page No. 1!

TITLE   NASHVILLE WITH A BULLET

PICTURE NO.

DATE PREPARED   May 1, 1980

| ACCOUNT NUMBER | DESCRIPTION | DAYS, WKS, OR QUANTITY | RATE | TOTALS | |
|---|---|---|---|---|---|
| 22 | GENERAL OVERHEAD | | | | |
| | A.   FLAT CHARGE | | | | |
| | B.   CORPORATE OVERHEAD EXPENSE | | | 51,000 | |
| | C.   CASTING OFFICE SALARIES | | | | |
| | D.   ENTERTAINMENT - EXECUTIVES | | | | |
| | E.   TRAVEL EXPENSE - EXECUTIVES | | | 2,000 | |
| | F.   OFFICE RENTAL AND EXPENSE | | | 2,000 | |
| | G.   AUDITOR | | | 2,200 | |
| | H.   TIMEKEEPER | | | | |
| | I.   SECRETARIES | | | | |
| | J.   PUBLIC RELATIONS HEAD | | | | |
| | K.   PUBLIC RELATIONS SECRETARY | | | | |
| | L.   LEGAL FEES | | | 20,000 | |
| | M.   OFFICE SUPPLIES | | | 1,000 | |
| | N.   POSTAGE - TELEPHONE & TELEGRAPH | | | 1,000 | |
| | O.   CUSTOMS BROKERAGE | | | | |
| | P.   CONTINGENCY ( COST OVERRUN FACTOR 15% of TOTAL | | | 590,595 | |
| | Q.   GENERAL OFFICE O.H. | | | 1,200 | |
| | R.   FILM SHIPPING | | | 500 | |
| | TOTAL GENERAL OVERHEAD          (without contingency) | | | 80,900 | 00 |
| | GRAND TOTAL | | | 671,495 | 00 |

ENTERPRISE STATIONERS  HOLLYWOOD

# 2 Managing People and Productions

In Chapter 1 we learned that producers must be successful salespeople. But once producers have sold a project, whether it is a pilot for a new television series or a 30-second commercial for a local retail store, they put on a brand new hat—they must become managers.

This job requires skill, tact, a firm knowledge of what motivates human beings to do things and a fine sense of organization. Whether one is managing a shoe store or is an independent producer, certain basic methods of management apply. We'll review some of these briefly in the first section of this chapter.

The independent producer's job, however, is also different from many other managerial positions and is somewhat unique. The producer walks a thin line between satisfying two disparate types of people. The client, as one type, is generally a businessperson, used to a certain set of stereotypical behavior patterns. The producer has to know how to manage those needs.

The other type includes those who work in the artistic and technical crafts, the cast and crew. They are of a different cut altogether. They are creative, and creative people often have special needs and require greater understanding. We will go into some detail on managing the production people later in this chapter.

In most businesses there is an axiom that states, "The customer is always right." The independent producer's customers, however, are the clients, and because clients are usually unfamiliar with the idiosyncrasies of video production, they cannot be expected to be right all the time. This chapter will examine effective ways of handling some of these delicate situations.

## BASIC MANAGEMENT IN REVIEW

Two primary theories of management are being taught in business schools today. Both discuss how to draw productivity from employees, but the means for achieving productivity are at variance. For the purposes of our overview, we will refer to the two schools of thought as traditionalist and behavioralist. In very simple terms, traditionalists manage by directive, while behavioralists manage by objective.

A traditionalist management makes decisions and passes them along to the workers in the form of orders and quotas. The system relies on rewarding extra output with bonuses and other pay incentives. Behavioralists put more emphasis on creating a positive environment and letting the workers feel that they have a say in the decision-making process. Employees are given the chance to set goals or objectives for themselves, rather than having them imposed by the manager.[1]

There is no one "best" way to manage. Producers must use elements of both techniques plus much creative "winging it," depending on their personality type. Some producers, for example, are involved in every step of the game from initial idea to the final sound mix. They are on the set giving advice and suggestions to the director, the sound technician and everyone else within earshot. Other producers are content to make the big decisions, hire experts for each of the key crew positions, and then get out of the way and let them all do their jobs. We'll discuss these two approaches later.

Regardless of what your approach to leadership will be, there are areas in which both the traditional and the behavioral methods of management agree. The following discussion coalesces these forms into some purely practical advice.

### The Human Touch

Most of us need not be psychoanalysts to understand the basic motivations of other people. We share, no matter what our station in life, a common humanity. The producer as manager must remember this simple fact above all others. While it seems that intelligence comes in different degrees, no one likes to be thought of as stupid.

The first rule of good management, then, is to treat your workers like human beings. Good leadership is always proven by example rather than dictate. If you expect your crew to be cheerful and easygoing, then you must be the same. If you expect others to work overtime, then you must set the example by being the first to come in early or stay late. You seldom have to go beyond your own instincts to know what to do with employees; insight is the key word. If possible, draw from your own experience; if you have ever worked under a management that treated its employees poorly, you know that this type of management style breeds contempt and non-productiveness.

## Communication

While a video production needs a spot at which "the buck stops," it remains a communal project. Everyone involved should have a chance to give you some input. Listening to your employees is not only good for their self-esteem, but may save you a good deal of grief down the line. No one person, even a great producer, is expected to be perfect or to have thought out all of the ramifications of a given situation. Many times employees are closer to the problem than you are. They can see what you might have missed. Therefore, good communication on the job is imperative.

Good communication from you to the employee means clarity, conciseness and a cordial delivery—whether written or spoken—so the employee knows what you expect. Good communication also means effective listening on your part. There are three types of listening: listening for information; listening critically (evaluating persuasive messages); and listening for feelings. Your employees will tell you quite a bit about how they are feeling—and therefore what you might be able to anticipate in terms of developing problems—if you give them the chance. And this means more than being a passive receptor.

Don't expect your employees to take the initiative; it's up to you to solicit employee input. You can do this in regular meetings or simply by dropping by a worker's desk and saying, "Frieda, would you have a look at this proposal and let me know what you think of it?" And when Frieda lets you know what she thinks, give her thoughts some real consideration. She may have spotted a flaw or come up with a new angle that you overlooked. If she hasn't, at least you have treated her as a valuable team member and have earned her respect. (See the Bibliography for a list of good books on the subject of communication.)

## The Good Word

Almost every person on this planet responds to praise. We seek approval, we need it and hardly any of us receive enough. If we do our job satisfactorily, we receive a paycheck, and many managers think that is enough. A paycheck may not be sufficient, however. Look inward to find the value of a pat on the back, a good word, a simple thank you for a task well done. The Academy Awards, the Emmy Awards, the Nobel Prize—all of these are nothing more than formal pats on the back. These awards share a common characteristic and one that is crucial to effective management—public praise.

If Frieda comes up with something you use in the proposal, be sure to tell everyone involved in the office. Reward her in public with sincere thanks for her help. This will do several things for the morale of your employees. First, they'll know that you're prepared to acknowledge their help. Second, they'll know that you do not regard yourself as infallible. Finally, it will spur them to work harder and be helpful themselves, and not to settle for a mediocre performance. The tough part about this area of management is that producers do not get any pats on the back because they are on the highest rung of the ladder.

## The Reprimand

From time to time, an employee makes an error. Sometimes the mistake is simply a bad judgment call. Sometimes it is due to a bad attitude. Whatever the cause, the producer/manager has to deal with it.

Since we all have grown up in a social milieu of reward and punishment, we expect that we will be reprimanded when we do something wrong. Many managers have the most difficult time of all disciplining an employee. They put it off, ignore the problem, redo the job themselves—anything to avoid a confrontation. This ostrich-like behavior is extremely bad for the team. Other employees will know that the error has not been acknowledged and morale may suffer. It's just as important for a manager to address a problem quickly and effectively as it is to reward good work in the same way.

The first rule of admonishment is the inverse of the first rule of praise. Always criticize an employee in private. Call the offender into your office or find a spot where only the two of you can be seen or heard. Remember that you are not dealing with a child, but another adult. Lay out your case in specific terms: "John, I just found out that the order for the light kits for the Jones Hotel shoot was not placed. Can you tell me what happened?"

By opening in this nonaggressive manner, you have cleared the way for John to give you a reason for the error without putting him on the defensive. He may open your eyes to a procedural problem of which you weren't aware with something like, "I couldn't place the order without a purchase request and accounting hasn't given me one yet."

If it's John's responsibility to do the ordering and he's having trouble with someone in another area, you need to know about it. You also must tell John that he should have notified you sooner about the situation if he couldn't handle it himself. If he just plain forgot to do the ordering, you need to let him know that you are not happy with his performance.

Whatever the cause of the situation, you have to remain calm and cool. Losing your temper, shouting or making threats is not productive. It is your job to keep the conversation rational.

Most employees will feel bad about causing or being part of a problem; therefore, assume that employees want to make amends. Turn this negative moment into a positive one by making employees feel as though they can be a part of the solution. Ask them for suggestions on how to rectify the situation and, more than likely, they will leave your office feeling chastened but positive about continuing the work.

This does not mean that you can just slough off the problem. Make employees aware of the seriousness of the error and the fact that you can't accept continued work of this quality from them. But you can, with some consideration for their feelings and a sublimation of your temper and/or frustration, turn them around to get on with the job.

## The Challenge

If you give your employees a specific set of tasks to accomplish in an orderly and timely manner, they will probably do just that. They will probably also determine that you don't expect them to do anything above and beyond those specifics. In order to reap the most from employees, an effective manager must learn how to present challenges. Most people fall into an inertial laziness if they don't feel challenged slightly beyond their abilities. They will become bored and complacent, neither of which is good for building a dynamic business.

The producer/manager walks an extremely fine line here. You can't overchallenge by setting goals that no one can be expected to accomplish. That will simply frustrate and ultimately destroy your staff.

What you must be able to do, however, is recognize the full potential of each employee by paying attention to what is going on. Encourage employees to move into other areas of a project, rather than directing them to more "busy work." If John, for example, has done all of his ordering and paperwork for the day and you find that he has time on his hands, you might encourage him to help Frieda in accounting. Or you might ask him to work on a new procedure for ordering.

It is important that employees develop a sense that nothing in your business is cast in concrete. Let them know that you are receptive to new suggestions and that you have established a policy of rewarding extra achievement. Those things alone present a challenge to anyone who wants to excel.

## Consistency

Nothing will destroy the spirit of a team working toward a common goal more quickly than inconsistent leadership; you must do everything possible to present a consistent image. If you come into the office one morning all smiles and good cheer and the next day show up looking like someone ate your porridge and set your house on fire, the staff will be on edge and not know what to expect from you. Everyone has an off day now and then. If you're feeling ill or depressed once in awhile, it simply means you're human. But if you make a habit of exhibiting extreme mood or personality swings, you will find yourself going through good people by the score.

You must treat all of your employees equally. It is quite common for a manager to like one employee better than another. If this turns into favoritism, however, it will spell disaster for your team spirit. If you fraternize with one of your employees, fraternize with them all. To maintain a professional business image and to retain a camaraderie on the team, the boss must be seen as a person who is fair, equitable and just to everyone.

### Delegate Authority

As a producer/manager, you are ultimately responsible for everything. You can't delegate that responsibility to anyone else. What you can and must delegate, however, is authority. You simply can't do it all. If you continually meddle in the jobs that have been assigned to your employees, looking over shoulders constantly, you will make your employees feel like robots.

From the stock boy to your associate producer, everybody likes to feel a degree of autonomy on the job. John likes to know that the ordering department is all his; he doesn't want to have to check with you on every detail. It gives him a feeling of satisfaction and worth to do it himself; to feel that he alone is responsible. Once he knows what your policy is and what you expect from him, leave him alone to do his job. The only time you should step in is when he makes a mistake. Even then, as we've discussed, you should try to let him solve the problem.

### Recognize Individual Needs

Finally, managers must recognize the uniqueness of individuals and realize that each employee has special needs. Some people require more positive reinforcement than others. Some require much structure on the job and appreciate specific task assignments. Other workers will be self-starters, needing very little motivation from you. It takes little more than a genuine interest in the welfare of your people to find these differences and to provide for them in your management.

## PRODUCTION MANAGEMENT

Production management is much more specific than general office management. Larger production companies take much of the load off the producer by hiring specialists in the field of production management. Most smaller companies cannot afford this luxury, however, and so it falls upon the producer to do the job of the unit manager and production manager for each project. All of this can be complicated if the producer fails to understand that he or she is dealing firsthand with a number of creative people.

These "creative types" tend to be a little more zany than their counterparts in nonproduction businesses. They tend to have a different view of themselves as employees than do people working on an automobile assembly line, for instance—not a better view, only different.

For the most part, "creative types" are exceptionally intelligent, artistically gifted humans with unusual senses of humor. They don't like to be "pushed," but will go to any length for the sake of the production once they feel as if they are part of it. While most of the management principles covered in this chapter certainly apply to production managers, they have to approach a creative crew expecting to be very flexible and willing to be tested.

The mechanics of production management is relatively straightforward and has been codified somewhat over the years. By breaking the production unit into departments, each with its own head, the production manager has a system for reporting and accounting.

## The Production Schedule

A team can function well only when each member knows what to do, when to do it and where to be. This information is derived from the production schedule, which includes estimates of how long it will take to prepare, shoot and complete the production. Because of the nature of creative projects, no schedule can be expected to survive intact from the moment it is created. Frequently, daily changes are made as unforeseen problems come up on the set. Scheduling, therefore, is a continuing process throughout the production, and this process must begin with the script.

### The Script Breakdown

The script has two major components: the scene description and the dialog or narration. We are most concerned with the scene description for our script breakdown. Figure 2.1 is a reproduction of the script for a 60-second television commercial for a restaurant entertainment complex that was still under construction at the time of the advertising campaign. The client wanted to stimulate interest in the complex before it was finished and encourage people to buy charter subscriptions to the private "key club."

Figure 2.2 is the same script written in split-page format. The process of production scheduling is the same regardless of the length of the script or its format. We will use this actual project as an example throughout the rest of the chapter.

Extracting Information from the Script Description

From the scene description, you should be able to extract the following information: which scene? who? what? where? when? props needed? costumes needed?

Figures 2.3 and 2.4 show the first page of our script marked to demonstrate how you would go through every page of your script to identify each element for the breakdown. Notice that we've drawn a line between each scene or numbered the scenes as in the split-page format. Each person, whether a major character or background extra, is circled. Essential props and costumes are underlined. Directors also circle and underline items to indicate other important considerations such as special-effects or lighting requirements.

**Figure 2.1: Fourdrinier Script**

---

**THE FILM AND VIDEO RANCH** 3674 Knapp Street Road   Oshkosh, WI 54901

---

Client: Como of Wisconsin                FOURDRINIER SCRIPT
        4321 W. College Avenue
        Appleton, WI 54911               60 second

---

INTERIOR— BERGSTROM PAPER COMPANY – FOURDRINIER MACHINE – DAY

ANGLE ON SPRAYER

We open on a deep gold mist. SNAPZOOM back to reveal the sprayer on the Fourdrinier papermaking machine. The wet pulp races away on its screen. Lines converge. On the SOUNDTRACK we HEAR FOURDRINIER JINGLE UP.

ANOTHER ANGLE

We see a ROLLER. The paper pulp whizzes around it lit with colored gels. GEARS are seen in the foreground. They turn. The machine shakes in rhythm to the music.

ANOTHER ANGLE

Another part of the machine. We see it through a tunnel. At the end of the tunnel is the FOURDRINIER LOGO SUPERED. The shaking of the machine is accentuated.

dissolve to

CU ACTOR

A handsome young exec-type. We SNAPZOOM back to reveal him walking between the machines which dwarf him with their size. He gestures toward one of them.

ACTOR

    THAT'S THE RHYTHM OF A REVOLUTION. A REVOLUTION WHICH BEGAN
    IN 1874 WHEN THE FOURDRINIER BROTHERS REVOLUTIONIZED THE
    PAPER INDUSTRY WITH THIS MACHINE TO TURN WOOD PULP INTO
    PAPER.

soft cut to

ANGLE ON PAPER ROLL – END OF MACHINE

It is coming off the machine. ACTOR enters frame, tears off a strip of the paper and we DOLLY IN FAST to the paper. On it is the FOURDRINIER LOGO.

ACTOR

    NOW, OVER A HUNDRED YEARS LATER, THERE'S A NEW FOURDRINIER

**Figure 2.2:  Fourdrinier Script in Split-Page Format**

---

**THE FILM AND VIDEO RANCH 3674 Knapp Street Road   Oshkosh, WI 54901**

---

Client: Como of Wisconsin
      4321 W. College Avenue
      Appleton, WI 54911

FOURDRINIER SCRIPT

60 second

---

| VIDEO | AUDIO |
|---|---|

---

1. INT. BERGSTROM PAPER CO.
We open on a deep gold mist. SNAPZOOM back to reveal sprayer on Fourdrinier paper machine. The wet pulp races away on its conveyor screen. Lines converge.

(MUSIC UP- FOURDRINIER JINGLE)

2. CU ROLLER
The paper pulp whizzes around the rol-ler lit with colored gels. Gears are seen F.G. turning. The machine shakes back and forth in rhythm to the tune.

3. ECU ANOTHER PIECE OF THE MACHINE
We see it through a tunnel. At the end of it is a SUPER of the Fourdrinier lo-go. The shaking is accentuated.

dissolve to

                          ACTOR

4. CU ACTOR
He is a handsome young exec type. We SNAPZOOM back to reveal him walking be tween the machines which dwarf him. He gestures toward them.

THAT'S THE RHYTHM OF  A REVOLUTION. A

REVOLUTION WHICH BEGAN IN 1874 WHEN

THE FOURDRINIER BROTHERS REVOLUTION-

IZED THE PAPER INDUSTRY WITH THIS

MACHINE TO TURN WOOD PULP INTO PA-

PER.

soft cut to

                          ACTOR

5. MS PAPER ROLL
It is coming off the end of the mach-ine. ACTOR walks into frame, tears a strip of paper off. FAST DOLLY into the paper which has on it the LOGO

NOW, OVER A HUNDRED YEARS LATER,

THERE'S A NEW FOURDRINIER REVOLU-

TION IN NORTHEAST WISCONSIN.

**Figure 2.3: Fourdrinier Script Marked for Breakdown**

---

**THE FILM AND VIDEO RANCH** 3674 Knapp Street Road  Oshkosh, WI 54901

---

Client: Como of Wisconsin                    FOURDRINIER SCRIPT
        4321 W. College Avenue
        Appleton, WI 54911                   60 second

---

INTERIOR- BERGSTROM PAPER COMPANY - FOURDRINIER MACHINE - DAY

ANGLE ON SPRAYER

We open on a deep gold mist. SNAPZOOM back to reveal the (sprayer) on the    *prime*
Fourdrinier (papermaking machine.) The wet pulp races away on its screen. Lines   *wide angle*
converge. On the SOUNDTRACK we HEAR FOURDRINIER JINGLE UP.    *lens*

---

ANOTHER ANGLE                                               *Check D.P.*
                                                    *for color suggestions*
We see a ROLLER. The paper pulp whizzes around it lit with (colored gels.) GEARS  *to make it*
are seen in the foreground. They turn. The machine shakes in rhythm to the    *dramatic*
music.

---

ANOTHER ANGLE

Another part of the machine. We see it through a tunnel. At the end of the    *check*
tunnel is the (FOURDRINIER LOGO) SUPERED. The shaking of the machine is    *w/music*
accentuated.    *First three scenes at same location*

---

dissolve to    *Same machine*
               *as Sc.#1*
CU (ACTOR)                   *check wardrobe*    *dolly+track*
                             *for contemporary "look"*
A handsome young exec-type. We SNAPZOOM back to reveal him walking between the
machines which dwarf him with their size. He gestures toward one of them.    *sync sound*

                              ACTOR

          THAT'S THE RHYTHM OF A REVOLUTION. A REVOLUTION WHICH BEGAN
          IN 1874 WHEN THE FOURDRINIER BROTHERS REVOLUTIONIZED THE
          PAPER INDUSTRY WITH THIS MACHINE TO TURN WOOD PULP INTO
          PAPER.

soft cut to

---

ANGLE ON PAPER ROLL - END OF MACHINE

It is coming off the machine. (ACTOR) enters frame, tears off a strip of the paper,
and we DOLLY IN FAST to the paper. On it is the (FOURDRINIER LOGO)
            *dolly +tracks*
*far end of*                 ACTOR                          *sync sound*
*building*
             NOW, OVER A HUNDRED YEARS LATER, THERE'S A NEW FOURDRINIER

   *Check D.P. for*
   *BIG area lighting*
   *requirements*

## Figure 2.4: Split-Page Fourdrinier Script Marked for Breakdown

**THE FILM AND VIDEO RANCH 3674 Knapp Street Road   Oshkosh, WI 54901**

Client: Como of Wisconsin
        4321 W. College Avenue
        Appleton, WI 54911

FOURDRINIER SCRIPT

60 second

| VIDEO | AUDIO |
|---|---|
| | |

1. INT. BERGSTROM PAPER CO.
We open on a deep gold mist. SNAPZOOM
back to reveal sprayer on Fourdrinier
paper machine. The wet pulp races away
on its conveyor screen. Lines converge.

(MUSIC UP- FOURDRINIER JINGLE)

*need prime wide angle lens*

*Same location*

2. CU ROLLER
The paper pulp whizzes around the roller lit with colored gels. Gears are
seen F.G. turning. The machine shakes
back and forth in rhythm to the tune.

*Check D.P. for color suggestions to make it dramatic*

3. ECU ANOTHER PIECE OF THE MACHINE
We see it through a tunnel. At the end
of it is a SUPER of the Fourdrinier logo. The shaking is accentuated.

*Check w/ music*

*— check wardrobe for contemporary "look"*

*dolly & track*

dissolve to

4. CU ACTOR
He is a handsome young exec type. We
SNAPZOOM back to reveal him walking-be
tween the machines which dwarf him. He
gestures toward them.

*Sync sound*

*Same machine cue sc. #1*

ACTOR

THAT'S THE RHYTHM OF  A REVOLUTION. A

REVOLUTION WHICH BEGAN IN 1874 WHEN

THE FOURDRINIER BROTHERS REVOLUTION-

IZED THE PAPER INDUSTRY WITH THIS

MACHINE TO TURN WOOD PULP INTO PA-

PER.

*dolly & track*

*Sar end OS building*

soft cut to

5. MS PAPER ROLL
It is coming off the end of the machine. ACTOR walks into frame, tears a
strip of paper off. FAST DOLLY into
the paper which has on it the LOGO

ACTOR

NOW, OVER A HUNDRED YEARS LATER,

THERE'S A NEW FOURDRINIER REVOLU-

TION IN NORTHEAST WISCONSIN.

*Check D.P. for BIG area lighting requirements*

*Sync sound*

In the margins we have made notes about other elements of the scene that we can infer even though they may not be called for specifically. The scriptwriter may not have specified each little detail that will be necessary to create the atmosphere of the scene. You have to fill them in here, based on your scouting of the location. After you have analyzed the entire script in this way, you are ready to move to the second phase of script breakdown.

Production Lists

On separate sheets of paper, one for each production element, make simple lists of the required pieces. Make one list for each category (sets, cast, props, costumes, details, etc.). Figures 2.5, 2.6 and 2.7 are examples of forms to help you make these lists.

A key element to remember in making up your set lists is that you will be scheduling all scenes shot on a certain set one after the other, even though they may appear at widely separated places in your script. For example, if the Fourdrinier spot ended at the paper mill, you would shoot the end scene immediately after shooting the beginning scene because you are in the proper location. This is called shooting out of sequence. So, when you're making your set list, you will list the set only once and then put in all of the scenes to be shot there.

The cast list has the names of the actors followed by all of the scenes in which they appear. The same is true of each of your other lists. The name of the prop, costume, etc., is followed by a list of all the scenes in which it will be needed.

The detail list, sometimes called an "insert list," is one you prepare in order to track small bits that can be recorded at one time without keeping the entire cast and crew standing around waiting. These details become very important in post-production when your editor is stuck trying to make a cut because of a mismatch in screen direction or a jump of some sort that would be jarring.

Inserts can also save you considerable grief if your piece is ending up too long and you need a place to get out of a long scene. In video news, these details or inserts are called "B-roll." Your editor will bless you for providing as many of them as possible.

Breakdown Sheets

Using your lists, fill in the blanks on each set's breakdown sheet (see Figure 1.5). Once you have done this you will have a document for each location, telling you how many pages of script will be shot there, the cast needed, the number of scenes, and details concerning props and wardrobe, effects, necessary construction materials and music playback. With all of this information categorized and filed in a looseleaf notebook to facilitate changes as they occur, you and your production staff are ready to move ahead.

**Figure 2.5: Property Form**

Reproduced with the permission of Don Kirk Entertainment Enterprises.

**Figure 2.6: Wardrobe Form**

# Wardrobe Record 8113

| COSTUME NO. | SCRIPT CHARACTER | ACTOR | DESCRIPTION OF COSTUME (INCLUDE ALL ACCESSORIES) | QUANTITY | SCENES | TECH. ADV. REQ. | FITTING REQ. | DATE FIRST FITTING | DATE SECOND FITTING | SOURCE NAME ADDRESS PHONE NUMBER | DATE READY FOR PROD. | STATUS† |
|---|---|---|---|---|---|---|---|---|---|---|---|---|
| | | | | | | Y / N | Y / N | | | | | |
| | | | | | | Y / N | Y / N | | | | | |
| | | | | | | Y / N | Y / N | | | | | |
| | | | | | | Y / N | Y / N | | | | | |
| | | | | | | Y / N | Y / N | | | | | |
| | | | | | | Y / N | Y / N | | | | | |

DON KIRK ENTERTAINMENT ENTERPRISES   RT. 9, BOX 127, CANYON LAKE, TEXAS 78130

† CODE: "S" - IN STORAGE   "P" - TO BE PICKED UP   "1" - MADE   "2" - PURCHASED   "3" - BORROWED   "4" - RENTED   "O" - OTHER

PRODUCTION _____
DIRECTOR _____
COSTUMER _____

PAGE ☐ OF ☐
DATE _____

Reproduced with the permission of Don Kirk Entertainment Enterprises.

**Figure 2.7: Makeup and Hairdressing Form**

# Makeup & Hairdressing Record 8131

PRODUCTION _____ NO. _____ DATE _____

| ACTOR'S NAME |
|---|
| Actor's Race or Ethnic group |
| Actor's Age and Sex |
| Special Facial and Hair Features |

| CHARACTER'S NAME |
|---|
| Race or Ethnic Group |
| Age and Sex |
| State of Health |
| Social Position or Wealth |
| Time Period |
| Season |

**SPECIAL SCRIPT REQUIREMENTS**
For Look of Character:

**SCENES or SHOTS** where character's makeup and/or Hairdressing is to be the same:

| MAKEUP MATERIALS |
|---|
| Foundation |
| Rouge |
| Eyeliners |
| Liners for modeling |
| highlight |
| shadow |
| Eyelashes |
| Eyebrows |
| Powder |
| Lipstick |
| Body makeup |

| HAIRDRESSING NEEDS |
|---|
| Applicants |
| Dyes |
| Cleanser |
| Wigs |
| Pin-up items |
| Trimming and Shaving |
| Style notes |
| Other hair items |

**SPECIAL EFFECTS**

**OTHER NOTES**

| | |
|---|---|
| Estimated PREP time: | Actual: |
| Estimated CLEAN time: | Actual: |
| HAIR photo taken: YES NO not yet No.: | |
| MAKEUP photo taken: YES NO ny No.: | |
| Hair to be done by: | Actual: |
| Makeup to be done by: | Actual: |
| This page COMPILED BY: | Date: |

DON KIRK ENTERTAINMENT ENTERPRISES   RT 9, BOX 127, CANYON LAKE, TEXAS 78130

Reproduced with the permission of Don Kirk Entertainment Enterprises.

*The Shooting Schedule*

Based on your experience with previous shoots, you can tell approximately how many hours you will need for each page of script. (If you haven't had a lot of experience, refer to the discussion of estimating time in Chapter 1.) Include the director and videographer when you compose the shooting schedule (see Figure 2.8) because they will be essential in making the schedule work.

Many directors, for instance, require much rehearsal time for their actors. Some do a great number of takes so they can select the best ones in post-production. Others pride themselves on spontaneity on the set, allowing the actors to improvise lines and situations. These considerations have an effect on the schedule.

The videographer, who sometimes is called director of photography (DP), also has a unique way of doing things. Some take a long time to decide on a lighting setup in order to make it perfect on the first take. Others, working with gaffers who seem to have an extra adrenal gland, delight in seeing how quickly they can achieve each setup. There is no right or wrong way, of course. For the producer, it is a matter of being able to work with creative people and allowing them as much freedom as the budget will permit. It is folly to create a shooting schedule that does not take into account the kinds of people who will be expected to follow it.

When the shooting schedule is completed, the producer finally has the documentation to do one other essential thing: predict the actual production budget that could be estimated only roughly with the data in Chapter 1. Going through your budget estimate again at this point will give you a very real approximation of costs.

It is wise to remember that usually the most expensive item in your budget is going to be talent. If, for example, major celebrities perform in your production, you may be paying them several thousand dollars a day. In that case, it would be better to make your shooting schedule fit around the time you have budgeted for your stars. Shoot all the scenes involving them, cut them loose from the payroll and then go back and shoot all the scenes using less expensive talent according to the one-set rule.

You can apply this same budget consideration if you have a particularly expensive piece of equipment to rent; a helicopter and Tyler Mount for instance. At $10,000 a day, you would obviously schedule all of your aerial sequences to be shot on the same day. Whether the most expensive item is people or equipment, strive to schedule it for the least amount of time.

A final note about your shooting schedule explains in part the rather large contingency factor that has to be programmed into any production. Unless you live and work exclusively in the Southwest, you must take the weather into account for exterior shooting. It is best to schedule all of the exteriors first, if possible. And always have a backup location for an interior set when you have scheduled exteriors. Nothing can be

**Figure 2.8: Shooting Schedule**

# Shooting Schedule 8119

DON KIRK ENTERTAINMENT ENTERPRISES   RT 9, BOX 127, CANYON LAKE, TEXAS 78130

| SCENE NO. | SCENE DESCRIPTION | PAGES | CAST | EXTRAS ANIMALS | COSTUMES / PROPS CARS / SPECIAL EFFECTS | SPECIAL INSTRUCTIONS |
|---|---|---|---|---|---|---|
| 3 | INT. NELSON HOUSE-DAY Jim and Peggy discuss the project. Laurie comes in from school. | 2 | JIM NELSON PEGGY NELSON LAURIE NELSON | BERENICE THE WONDER DOG | JIM: cross country ski outfit. PEGGY: wool ski sweater, brown pants, apres ski boots. LAURIE: blazer, wool skirt, snow jog boots. | Check on artificial snow with Mazzio Supplies on 2/7. Crew call 6:00a.m Cast call 9:00a.m Director 9:00a.m Arrange hot lunch for cast & crew. Note: Laurie is on salt free diet |
| 8 | INT. NELSON HOUSE-NIGHT Jim and Peggy preparing to go to awards banquet as Laurie does homework | 4 | SAME | SAME | PROPS: Computer terminal, household dressings. Two pair cc skis and shoes. Backpack and school books. FX: Snowfall visible through L.R. bay window. SC. 8: JIM: Business suit. PEGGY: Givenchy dress. LAURIE: P.J.'s & robe. Night lighting. Snowfall. | SC.8. Light for night effect. Arrange hot supper for cast & crew. Peggy must have fish only. |

LOCATION / ADDRESS / CONTACT / PHONE
3674 Knapp Street Road/Max Romanowski/424-3134

FILM TITLE The Mechanisms of Joy    WEEK [2]

DAY [4]  DATE 2/9/86    PAGE [1/2-4]

Reproduced with the permission of Don Kirk Entertainment Enterprises.

more frustrating—and more expensive—than watching a cast and crew scheduled for a sunny exterior stand around under umbrellas waiting for the rain or snow to stop!

## Pre-production Planning

At this point, all of your department heads are called together to begin pre-production planning, the process most frequently overlooked or underdone by the novice. It is, however, recognized by the seasoned professional as the heart and soul of the operation.

Pre-production planning, called prepro in the jargon of the industry, can make the production run like a precision machine. This is where you and your department heads do all of the organization. You will have daily production meetings from this point on through production (called the shooting phase or principal photography).

The following department heads should be present at the production meetings: directing, art, camera, sound, makeup and costumes, transportation and, of course, production. Although one person may have to do more than one job in very small operations, the duties of each department head are described below.

### Directing

The director's staff consists of the director and at least one assistant director (AD). The function of the first AD, curiously, has very little to do with directing. He or she is really more of a producer's functionary, acting as liaison between the director and the producer, the director and the technical crew, the director and virtually all problems that could distract him or her from the creative task of bringing the script to life on tape.

First ADs are responsible for revising the shooting schedule as changing requirements dictate. They make up a Daily Call Sheet (Figure 2.9) that lists everybody and everything needed for the day's shoot, along with information on the time and place where everybody and everything should be. First ADs also make and distribute daily work orders, which list all the technical requirements for each day.

Daily Production Reports (see Figure 2.10) are also made and compiled by the first ADs. From these reports, they put together the Daily Log (Figure 2.11), which is a summary of each element of each day's shooting. From this form the producer can make exact records of times and costs, which will be of tremendous assistance when budgeting for succeeding productions.

During the prepro phase of your project, the first AD is preparing all of these forms while doing everything possible to meet demands for help from the director. A good first AD is the second most important member of the team. This position also usually deserves a helper called the second assistant director to do much of the actual leg work.

**Figure 2.9: Daily Call Sheet**

# Call Sheet 8103

_____ DAY OF SHOOTING
DATE _____
CREW CALL _____
SHOOTING (LV) CALL _____

PRODUCTION _____ NO. _____
PRODUCER _____ Asst. DIR. _____
DIRECTOR _____ UNIT PROD. MGR. _____
COVER SET:
CREW Report to:

SUNRISE _____
SUNSET _____
DAILIES:

DON KIRK ENTERTAINMENT ENTERPRISES   RT 9 BOX 127, CANYON LAKE, TEXAS 78130

| SET | SCENES | PAGES | CAST NOS. | LOCATION |
|-----|--------|-------|-----------|----------|
| | | | | |

| NO. | CAST & DAY PLAYERS | PART OF | MAKEUP | SET CALL | REMARKS |
|-----|--------------------|---------|--------|----------|---------|
| | | | | | |

| ATMOSPHERE & STANDINS | PROPS | SPECIAL INSTRUCTIONS |
|-----------------------|-------|----------------------|
| | | |

| NO. | ITEM | TIME | NO. | ITEM | TIME | NO. | ITEM | TIME | NO. | VEHICLES | TIME |
|-----|------|------|-----|------|------|-----|------|------|-----|----------|------|
| | DIRECTOR | | | GRIPS | | | TEAMSTER | | | | |
| | FIRST A.D. | | | ELECTRIC | | | WRANGLER | | | | |
| | SECOND A.D. | | | ART DEPT. | | | SECURITY | | | | |
| | PROD. ASST. | | | PROPERTY | | | STILLMAN | | | | |
| | CRAFTSERV. | | | LOCATION | | | | | | | |
| | SCRIPT SUPR. | | | SPECL. EFX | | | | | | | |
| | DIR. of PHOTO. | | | MAKEUP/HAIR | | | | | | | |
| | CAMERA | | | SFX MAKEUP | | | BREAKFAST | | | | |
| | SOUND | | | WARDROBE | | | LUNCH | | | | |

ADVANCE SCHEDULE OR CHANGES

Reproduced with the permission of Don Kirk Entertainment Enterprises.

**Figure 2.10: Daily Production Report**

# Daily Production Report 8104

PRODUCTION _____ NO. _____ DATE _____

DIRECTOR _____

DIR. of PHOTOGRAPHY_____

CAMERA OPERATOR _____

SOUND RECORDIST_____

DATE PROD. STARTED _____ EST. FINISH DATES _____

| NO. of DAYS on PICTURE | | | | |
|---|---|---|---|---|
| TRAVEL | REHEARSALS | IDLE | WORK | TOTAL |
| | | | | |

SET_____ LOCATION _____

COMPANY CALLED_____ FIRST SHOT_____ LUNCH: from_____till_____ CAMERA WRAP_____

| | SCENES | PAGES | MINS. | SETUPS | TAKES | ADDED SCENES | RETAKES | STILLS |
|---|---|---|---|---|---|---|---|---|
| PREVIOUS | | | | | | | | |
| TODAY | | | | | | | | |
| TOTAL | | | | | | | | |

| SCENE NO. | | | | | | | | | |
|---|---|---|---|---|---|---|---|---|---|
| ADDITIONAL SCENES | | | | | | | | | |
| RETAKES | | | | | | | | | |
| SOUND TRACKS | | | | | CREDITS | | | | |

| PICTURE NEGATIVE | | | VIDEO TAPE | | | RECORDING TAPE | | |
|---|---|---|---|---|---|---|---|---|
| PREVIOUS | TODAY | TO DATE | PREVIOUS | TODAY | TO DATE | PREVIOUS | TODAY | TO DATE |
| ON HAND: | | | ON HAND: | | | ON HAND: | | |

| NEGATIVE BREAKDOWN | | | | REMARKS AND EXPLANATION OF DELAYS | ESTIMATED DAYS ON PRODUCTION |
|---|---|---|---|---|---|
| | GOOD | WASTE | DRAWN | | |
| PREVIOUS | | | | | |
| TODAY | | | | | TOTAL SCENES IN SCRIPT |
| TO DATE | | | | | |

| CAST | STATUS | CALL | WRAP | LUNCH | HOURS | EXTRAS/STUNTMEN | | | TOTAL PAGES |
|---|---|---|---|---|---|---|---|---|---|
| | | | | | | NO. | STRAIGHT | OVERTIME | |
| | | | | | | | | | TODAY'S WEATHER |
| | | | | | | | | | |
| | | | | | | | | | |
| | | | | | | | | | |
| | | | | | | | | | |
| | | | | | | | | | |

ADVANCE SCHEDULE:

DATE_____SET_____TIME CALLED_____LOCATION _____

REMARKS _____

PRODUCTION MANAGER _____ ASSISTANT DIRECTOR _____

DON KIRK ENTERTAINMENT ENTERPRISES  RT 9, BOX 127, CANYON LAKE, TEXAS 78130

Reproduced with the permission of Don Kirk Entertainment Enterprises.

**Figure 2.11: Daily Log for Fourdrinier Script**

```
DAILY LOG                      Production F&VR-012-7 FOURDRINIER
Crew Call -9:00a.m.   Director- 9:00a.m.    Shooting -10:00a.m.

Sc 2-1      9:00-9:40     Setup equipt. Lineup & light INT. BERGSTROM
            9:40-9:55     Place actor & background workers
            9:55-10:04    Director walks through w/talent
           10:04-10:17    Adjust lighting & practice dolly
           10:17-10:25    Shoot 1 take
           10:25-10:30    Shoot 2 takes
   3-1     10:30-11:00    Move setup down the machine & relight
           11:00-11:03    Shoot 1 take
           11:03-11:06    Shoot 2 takes
   3a-1    11:06-11:25    Blown circuit; rewire & run cable
           11:25-11:30    Shoot 1 take
           11:30-11-42    Adjust lights
           11:42-11:45    Shoot 2 takes
   4-1     11:45-12:02    Relocate down the machine
           12:02-12:22    Adjust light & practice camera move
           12:22-12:28    Shoot 1 take
           12:28-12:50    Blown circuit; re-rig & relight
           12:50-12:55    Shoot 2 takes
           12:55-1:10     Director decides on another shot. Move
                          camera to top of machine.
            1:10-1:15     Shoot 1 take
            1:15-1:20     Shoot 2 takes
            1:20-2:20     Lunch at Valley Inn across street
            2:20-3:20     Shift all equipment to other end of factory
                          relight and lineup
   5-1      3:20-4:00     Director explains shot. Rehearse w/dolly.
                          Humidity causes lens fog. Hair dryer obtained
                          to clear fog during shot.
            4:00-4:05     Shoot 1 take
            4:05-4:22     Actor having trouble making move. Rerig
                          lights and dolly track.
            4:22-4:30     Rehearse new move
            4:30-4:34     Shoot 1 take
            4:34-4:40     Adjust lights
            4:40-4:44     Shoot 2 takes
   6-1      4:44-5:00     Move setup to end of machine
            5:00-5:30     Relight and practice fast dolly
            5:30-5:45     Rehearse talent w/camera move
            5:45-5:46     Shoot 1 take
            5:46-5:55     Adjust lights.
            5:55-5:56     Shoot 1 take
            5:56-6:00     Rehearse move w/actor
            6:00-6:12     Shoot 4 takes
            6:12-6:45     Shoot inserts-4 takes
            6:45-7:12     Strike set & equipment
            7:12          Company dismissed w/ meal penalty
```

During prepro the director is casting talent, scouting locations, planning camera shots, rehearsing talent (whenever the budget will allow for this luxury), discussing camera and lighting requirements with the DP and working closely with the art department.

## Art

During pre-production planning, the art department works on the storyboard with the director. Figure 2.12 illustrates a good storyboard format. Each major scene is broken down into drawings showing camera position and the major action to be included in the frame. The space below the drawing contains captions from the dialog or narration and notes upon which the art director and director have decided. If you can't afford a real artist to do your storyboard, do it yourself with stick figures if nothing else. It is a fundamental tool in any professional production.

The storyboard simplifies the instructions given to the camera and lighting crew, the actors and everyone else who is involved on the set. It also solidifies the director's thinking and gives the client, if there is one, a clear idea of what to expect on the screen.

Storyboards are guideposts at the outset of the production. The drawings are not indelible images impossible to be changed should the director see something better at the time of the shoot. But without this blueprint from which to begin, each scene, each day's shooting can be a nightmare of trying to decide what to shoot while your budget trickles away by the hour.

The art department is also responsible for such practical things as designing and building sets, dressing (rearranging and decorating) existing locations that are going to be used as sets during the shoot and coordinating literally all the visual elements with the director and the DP.

## Camera

The DP is in charge of the camera department and is ultimately responsible for all of the technical crew aside from sound. This includes the grips (whose foreman is the key grip), the gaffer and all who work for them. In a large-scale operation, this is what the minimum camera crew consists of:

1. Director of photography: designs the lighting setups, chooses the camera angles and lenses, coordinates with the director and supervises the technical crew.
2. First camera operator: operates the camera during the shoot.
3. First assistant camera operator: assists the operator in all functions, including taking distance measurements from camera to subject, setting up the camera, taking light measurements and setting the correct f-stop, pulling focus or zoom when necessary and making camera reports.

4. Gaffer: implements the instructions of the DP, rigs the lights and wiring on the set, patches into external power sources as necessary and is directly responsible for all the electrical wiring on the set.
5. Key grip: supervises the grip crew, does all of the practical moving of equipment for the shoot and puts lighting and lighting control instruments in place at the direction of the gaffer.

During prepro the camera department is very involved with locating all of the items that will be needed for production. Since most small production companies do not own as much equipment as is often necessary, the DP and producer should have a line on all the equipment rental houses. The producer should go through the catalogs of equipment with the DP and choose items based on expectations of need drawn from the script and storyboard.

Normally, you can and should trust the suggestions of your DP in technical matters. A wise producer, however, will have at least a speaking familiarity with the technical end of the production. You won't have to know what a "pigtail" or a "brute" are; an "octopus" and a "cookie" may remain in your mind as an ugly fish with eight arms and a flat, sweet thing with chocolate chips. But you should certainly know the difference between a three-tube or three-chip and a single-tube video camera; between top-of-the-line industrial and bottom-of-the-line broadcast-quality equipment.

You should understand the elementary components of a grip truck to be able to appreciate whether or not you will need an external generator on the set to handle the amperage of the lights you'll be using. There is a tendency for many technicians to want to use the latest and best when frequently the older and cheaper will do. Your DP must understand your budget limitations, so hold nothing back in your prepro discussions.

### Sound

In most smaller video productions, the VTR operator also monitors and adjusts the audio levels and is, in effect, the sound recordist. This can work when the project requires only one actor wearing a lavaliere microphone. If the sound recording is more complicated, however, one person is not sufficient.

Consider sound an integral department of its own, and staff it with a recordist and a boom operator. The recordist records the audio track on the VTR as well as on a backup audiotape deck and does on-location mixing of various audio sources, including multiple microphones, to provide daily sound reports. The boom operator places microphones on the set and on the actors. When unidirectional (shotgun) or other kinds of microphones held overhead on a boom or "fishpole" are used, the boom operator physically holds the pole and directs the microphone at the actors.

Your recordist as department head, while only minimally involved in prepro, must be given the chance to estimate the amount and kinds of microphones, cables,

**Figure 2.12: Storyboard Format**

# Storyboard 8128

SCENE NO. 7

**PRODUCTION** *FOURDRINIER*         **NO.** *80-03*   **DATE** *11-17-80*

DIRECTOR *Bob Jacobs* ARTIST *Same*       SCRIPT DATED *10-30-80* PAGE *2* OF *3*

SETUP NO. *27* SCRIPT PAGE NO. *2*

DIALOG AND/OR ACTION:
... AND THIS WINE AND CHEESE CELLAR ...

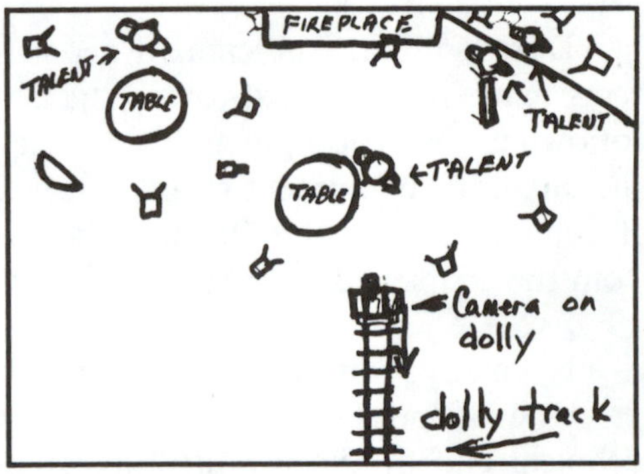

CAMERA PLACEMENT ON SET PLAN

START C.U.
Dolly back to reveal the entire room. LIGHTS SUSPENDED ON GRID.

SETUP NO. *27* SCRIPT PAGE NO. *2*

DIALOG AND/OR ACTION:
... TO GIVE YOU BOTH AN INTIMATE CONVERSATION PLACE OF YOUR OWN ...

CAMERA PLACEMENT ON SET PLAN

SAME

DON KIRK ENTERTAINMENT ENTERPRISES   RT 9, BOX 127, CANYON LAKE, TEXAS 78130

Reproduced with the permission of Don Kirk Entertainment Enterprises.

tape, etc., that will be required for the shoot. Many free-lance recordists come with their own equipment included in their daily or weekly rate. As a rule of thumb, remember this: If what is being said on the set by your actors is important, don't trust the recording to an assistant cameraperson or VTR operator—hire an expert.

## Makeup and Costumes

Even if you will be using "real people"—such as the actual maids in our Jones Hotel training tape—you will be wise to have a costume and makeup coordinator. On smaller shoots, one person can frequently handle both jobs. This department is responsible for obtaining and applying makeup; buying or making clothing; taking measurements of the cast members; providing a sufficient number of identical costumes so that clothing can be changed if soiled or rumpled on the set; and attending to all the details that will go into the look of your actors.

On major shoots, this department must sometimes begin work several weeks in advance of the actual production. In those cases you will have a head costumer, an assistant costumer, a men's dresser, a women's dresser, a hairstylist, a head makeup person, an assistant makeup person and, in the case of some productions, a men's and a women's body makeup person!

## Transportation

The transportation captain is responsible for arranging for and providing all movement of people and things during your production. This person hires trucks and other vehicles to ensure the daily pickup and delivery of your talent to and from the set, arranges airline reservations for your cast and crew as necessary, etc.

He or she will need your shooting schedule as far in advance as possible in order to carry out this heady responsibility. Most production companies rent all the necessary vehicles. If you are going to be on a remote location overnight, the transportation captain will also have to provide trailers or motor homes as dressing and makeup rooms for the talent, as well as possibly for the cast and crew to use as sleeping quarters.

## Production Office

During this phase of the operation you and your assistants are coordinating all the details from the information provided by your department heads. Having approved, for example, the game plan of your DP, you will be ordering the equipment from the rental houses, paying the bills, maintaining accounting records, etc. A good checklist for you at this stage is the Unit Manager's Worksheet (see Figure 2.13); this form is used daily during production.

On the worksheet you'll find practically everything you will need to think about and plan for. If you're shooting exteriors in any major city, for instance, you will need

**Figure 2.13: Unit Manager's Daily Worksheet**

# Unit Manager's worksheet 8121

**PRODUCTION** _____ **NO.** _____ **DATE** _____

☐ FIRST UNIT   ☐ SECOND UNIT   DIRECTOR _____ UNIT MANAGER _____

**SET** _____ **NO.** _____ **SCENE NOS.** _____

**LOCATION ADDRESS** _____ **CONTACT** _____ **PHONE** _____

☐ Rain or shine   ☐ Weather permitting   ☐ EMERGENCY NO.:
☐ ALTERNATE PLANS:

Expected Days
Weather:

**TRANSPORTATION PLANS**    Transportation Captain _____ ___ cars
☐ CAST:                                               ___ busses
☐ CREW:                                               ___ trucks
☐ EQUIPMENT:                                          ___ drivers

**PHONE NOS.**
Prod. Off. _____
Hospital _____
Doctor _____
Taxi _____
Police _____
_____
_____

**MEALS** no. of:
___ Staff
___ Cast
___ Crew
___ Extras
___ Drivers
___ Local help
___ TOTAL

**FOOD**    Qty.
☐ Coffee    ___
☐ Donuts    ___
☐ Hot Chocolate    ___
☐ Soft Drinks    ___
☐ Snacks
☐ Cups, Plates, Napkins, Silverware ___
☐ Refrigeration Required _____

**LUNCH**
☐ Packed Lunch
☐ Catered
☐ Restaurant
☐ Bring their own
☐

☐ TIME AND PLACE:

☐ SPECIAL CREW REQUIREMENTS:
VEGETARIANS no. ___

**REQUIRED FACILITIES**    LOCATION / NOTES
☐ Restrooms
☐ Dressing Rooms
☐ Rest Area
☐ Wardrobe Space & Racks
☐ Makeup Room & Tables
☐ Projection Room
☐ Tables & Chairs
☐ Overnight Accommodations
☐

**COMMUNICATIONS**
☐ Telephone
☐ Walkie-Talkie
☐ P.A. System
☐ Radio contact w/ studio

**FIRST AID**
☐ First Aid Kit
☐ Nurse
☐ Doctor
☐ Ambulance

**CROWD CONTROL**
PLAN:

**TRAFFIC CONTROL**
PLAN:

**CHECK ON**
Licenses ☐
Police Permits ☐
Additional AD's ☐
Identification Cards ☐
Insurance & Taxes ☐
Child Labor Laws ☐
Airline Schedules ☐

**MISCELLANEOUS**
☐ Route Maps issued
☐ Call Sheets distributed
☐ Releases signed
☐ Time Cards filled out
☐ Expenses logged
☐ Film & Reports shipped
☐
☐ Film Crates & Labels
☐ Production Forms
☐ Extra Film Scripts
☐
☐ Electrical Hookups
☐ Worklights needed

**ANIMAL CARE**
☐ Cages   ☐ Wrangler
☐ Food   ☐ Restroom

**PERSONNEL**    NAME    PHONE    DO ON SET    NOTES
☐ Child Attendant
☐ Interpreter
☐ Wrangler
☐ Security
☐ Special Effects
☐ Drivers
☐ Cleanup Crew
☐ Unit Publicist
☐

DON KIRK ENTERTAINMENT ENTERPRISES   RT. 9. BOX 127. CANYON LAKE. TEXAS   78130

Reproduced with the permission of Don Kirk Entertainment Enterprises.

to know whether or not the local police require you to obtain permits or licenses before shooting in public places.

Whenever a camera crew shows up at a location like a city street or a park, it is likely that the activity will attract a crowd. Either you or the police will have to provide crowd control in those circumstances. For that reason, many cities have established flat fees for the commercial use of public facilities—both interior and exterior. Fines for shooting without a permit are frequently rather steep. It pays, therefore, to make a call to your location's City Hall to check on this kind of detail before an oversight pops up to ruin your otherwise perfect day!

The production office also arranges for meals for the cast and crew. Generally, meals are brought to the shooting site to reduce downtime. Remember to find out if any cast or crew members have special dietary needs, and plan accordingly.

The production office coordinates everything that happens during pre-production and production; it brings together all the details to make the project work. A first-class secretary in this office is essential.

As the project moves from prepro into principal photography, your planning and management should make the transition as smooth as possible. You have instilled in your department heads a feeling of camaraderie and they know they have your total support. The pressure on you, however, doesn't let up for a second. The success or failure of the project and the return of a good product to your client is completely your responsibility.

Universal Studios' producer/director Joel Schumacher once stated some very practical observations about producing and the responsibilities that accompany it:

> My dream was that I would go on and make productions and have this wonderful life. And it is a wonderful life. And I wouldn't trade it for anything in the world. But, it isn't what I thought it was going to be. It's a people business. It's an ego business. You have to keep on going back in there when you maybe don't want to. And you have to keep on working on days when you don't feel like it and you have to work with people that . . . sometimes you don't like. They're not kind and they're not supportive and you still have to get the best out of them for the project whether you want to be there or not. That's what it means to be the guy at the top. And that takes integrity.[2]

The producer has to be able to keep three things in mind at all times—the project, the project and the project! A producer uses what wags refer to in the business as O.P.M. (i.e., other people's money). And when you're responsible for what happens with thousands and sometimes millions of dollars worth of O.P.M., you had better have plenty of integrity.

There's an old saying that goes, "It's lonely at the top." Being in business for yourself will prove that expression to be true. There is no one to give you the pats on the back you may need. Instead, you are expected to give them to everyone around you. As the boss, you are expected to be a fountainhead of strength, justice and support. At best, you will be second-guessed quite a bit; at worst, employees may ridicule you behind your back.

You will seldom, if ever, be fully understood, sympathized with or complimented by those who work for you. This is a natural condition in most employer-employee relationships. If you are superb at your job and manage to find exceptional people, you may establish the close "family" relationship that some producers enjoy. You must be prepared, however, to work within the more traditional framework.

Thus, you must be able to draw on your own wellsprings of satisfaction, be able to thrive and grow under constant pressure from all sides, make decisions quickly, work with and inspire other people by your own example, and finally, be able to derive pleasure from doing these things.

## MANAGING THE CLIENT

Whether your client is NBC or Jones Hotels, you will have creative control over the project. Stephen J. Cannell, a legendary producer behind a seemingly endless string of hit network shows such as "The Rockford Files," "The A-Team" and "Riptide," has some insight into dealing with clients:

> Network executives are for the most part businesspeople. They are accountants or MBAs. Now and then I get into, well, "discussions" with them about a creative element in one of my shows. Now if it's "Standards and Practices" or a thing like that, well, then they will have the final say. But if it's a matter of something I feel strongly about, like the characterization of this or that guy or the casting of this or that person, then I will try everything possible to have my own way. You have to remember that it is their money but if they've hired me to do what I do creatively, then I want them to let me do my job.[3]

Some clients hire you and leave you completely alone to do your job. More frequently, however, clients feel they should have input into the way things are done because they have put up the money.

Naturally, clients have script approval. It is, after all, their story that you are telling. At this stage you should be prepared to accede to reasonable requests for the insertion of this or that sequence, dialog and so on. You must also be able to argue for "the right and true" if you feel that the client is completely off base. Once you have hammered out the script and the clients have given their input, it becomes imperative that you be able to tell them, very diplomatically of course, to buzz off.

If your client shows up on the set every day, begins to interfere with the workings of your crew, starts telling the director what to shoot and how to shoot it and wants to put family members in the scene, then you are in for trouble. This is when your integrity will be tested.

If you don't care what your product looks like—if you're only in it for the money—then be very sure that your cast and crew know it. Everybody can humor the client and walk away happy at the end.

If you do care, then you will have to be fair, courteous but absolutely strong. Take your client to a private place and explain something like the following: "Mr. Jones, this shoot is costing you $5000 a day. I have your signature approving the script and the storyboard and I've budgeted this thing down to the last penny based on that approval. Now if you're going to keep insisting on changes and interfering on the set like this, we're going to have to do a whole new budget. I know that you mean well, sir, but we *are* the professionals on the production end and I'm going to have to ask you to let us do our jobs for you."

You can only hope that clients will be reasonable. Many clients and agency representatives actually contribute on the set by giving you fresh eyes and ears. Logic and tact will win the day for both of you in workable situations like this. But in extreme instances, rather than work yourself and your production people through a living nightmare, you must be prepared to either close down the project (returning the unused portion of the budget to the client with your best wishes), or grin and bear it through to the end. This decision will be the ultimate test of your integrity.

## SUMMARY

Effective management is a creative process. It requires an awareness of human motivations and a love of working with people. Management also requires a keen mind for details and an ability to anticipate and organize. We hope that the forms and procedures outlined in this chapter will help producers who wish to manage their productions, staff and clients with skill and effectiveness.

## NOTES

1. For a good book on the behavioralist technique, see Kenneth Blanchard and Spencer Johnson, *The One Minute Manager* (New York: Berkley Books, 1981). Other good books on basic management principles are listed in the bibliography at the end of this book.
2. Joel Schumacher, interview with author, January, 1982.
3. Stephen J. Cannell, interview with author, January, 1984.

# 3 The Independent Business: Setting Up Shop

The first two chapters of this book consider information that would be of use to all producers, whether they work for a major corporation in-house or for an agency or as a supplier to one of the networks.

This chapter is for producers who have made the decision, or are thinking about making the decision, to set out on their own. Deciding to go into business for oneself, to become the symbol of free enterprise—the entrepreneur, the small business-person—is one of the most crucial moments in a person's life.

## MAKING THE DECISION TO GO SOLO

If you are at this crossroad you must consider several factors. You will be giving up security, regular paychecks, fringe benefits, retirement plans, etc., along with the nicety of having someone else worry about meeting the payroll.

There are a number of reasons why people jump into their own business. Certainly, there is some glamour in pursuing the Great American Dream. But there are also some very real requirements for these special people. The first part of this chapter provides food for thought about joining the ranks of the self-employed.

### Entrepreneurship

An entrepreneur has to be a cross between a starry-eyed dreamer and a fundamental conservative capitalist. The dream part has to be so strong, so pervasive and such a driving force that the individual cannot bear to do anything else for it may have to sustain that person through disillusionment and hard times.

## The Bad News

The statistics regarding entrepreneurship are sobering. Only 67% of the 500,000 businesses a year that start up in America make it through their first year. Only 33% make it to five years and a mere 20% hang on for a decade.[1]

The primary reason for this failure rate, according to creditors and business analysts, is ineffective management. The U.S. Government's Small Business Administration (SBA) gives some advice in the next section about the kind of person who can get a business started and make it work.

## Assess Your Motivation

The SBA suggests that you think about why you want to own your own business and be certain that you want it badly enough to work long hours without knowing how much money you will make. Past experience—especially as a manager—in a business like the one you wish to start can prove very helpful; and assets such as formal business training and extra money cannot hurt.

Uncle Sam wants you to make it in business. And he can provide some help as we will see later in this chapter. But before you jump into the ocean of uncertainty, it is important to think about the kind of person you really are. According to the SBA,[2] these are the qualitites that make for a successful entrepreneur—match yourself up for comparison:

1. You are a self-starter. You do things on your own. Nobody has to tell you to get moving.
2. You actively like other people. You can get along with just about anybody.
3. You can lead people; get them to go along with your ideas and your momentum once you get started.
4. You enjoy responsibility. You seek it out, take charge of things and always see them through to completion.
5. You are a gifted organizer. You have a plan before you start a project and you are usually the one to get things lined up when the group wants to do something.
6. As a worker you can keep going as long as you need to. You never mind working hard for something you want.
7. You can make decisions in a hurry if you have to and they usually turn out to be right. Not perfect, just right.
8. Your word is your bond. You can always be trusted and never say things you don't mean.
9. You have persistence. You stick to any job you've made up your mind to do and you don't let anything stop you.
10. You are in top physical condition and you never get run-down.

These are assessments by the SBA of personality types who are likely to make it in business for themselves. Notice that almost every one of these applies to the producer-type we have discussed in previous chapters.

The conclusion that can be made here is rather simple. If you are currently a successful producer working for someone else and you have the drive and the capitalization, you have a good chance of succeeding on your own. However, you must assess your limitations as well as your motivation before you decide to go solo.

## Know Your Limitations

The following section is about limitations. It is also about common sense—an essential attribute for those who wish to understand limitations.

Caught up in the exhilaration of the moment, sometimes blinded by the light of our own derring-do, we can forget some basics—e.g., paying attention to details, remembering who we are and where we came from and why we went into business for ourselves. Some of the information and advice that follows is derived from the experience of people who learned their lessons the hard way—by going broke.

In a blind panic, with the monthly bills coming due again, when one or two key sales that will put you over the hump are just hanging fire, you may be tempted to deal with one of the thousands of idiots who hover about the fringes of this business. Be prepared for them to pester you. They seem quite legitimate on first approach. If you make the error of taking a chance on them, they will break you.

"They" are the dream merchants. They always have a million-dollar idea for a series, a made-for-TV movie, a special or a hot new toy to sell with late night, per-inquiry spot TV commercials. And they've chosen to cut *you* in on it. All you have to do is join them in their madness and they'll make you rich. Here are some examples of what happened when otherwise sensible producers forgot the main limitation involved here—time.

Your time is precious. Don't waste it as these producers did. The most serious mistake a new producer can make when venturing out into the world of independent production is overextension.

The names in these case studies have been changed; otherwise the stories are true.

### Case 1

The producer (Frank) was making good progress toward his goals. He had done work for a major corporation and a long-term contract was in the offing. He had taken on a partner who was good at sales. The future looked promising although accounts receivable were not actually flowing over.

Through an acquaintance of his sales partner, Harry, Frank met a prospect. The man had a small publishing business. Moderately successful at putting out a few do-it-yourself books and vanity novels, this man we'll call John had a dream.

He had a wonderful Christmas story that he wanted to do as a TV special. John told Frank and Harry that he had backing to the tune of $1 million. All he needed from them were their production skills and access to some of their contacts in Hollywood. The script was being written, he assured them, by a top-notch screenwriter. Did they want in?

Now ask yourself, given the above, if you would want in.

Frank and Harry did. And they ended up devoting nearly a year of their precious time to a project that never materialized. It was not a deliberate ruse, of course. John believed, as such people usually do, that saying something makes it so. He was in love with the story and knew other people would be, too. A friend of a friend had, indeed, mentioned something about being able to come up with $1 million for the right project with the right cast at the right time. John simply had faith that it would all come together because he wanted it to happen so badly.

Here's what really went on. The screenwriter turned out to be John's brother in St. Louis who thought he was as good a writer as those "hacks on TV." The story had been "borrowed" (without the publisher's permission) from a national magazine. Nobody had the rights to do it as a screenplay. The money guy turned out to be a middle-class stockbroker with an itch to break into Hollywood and enough personal debts to sink him at any moment. There was never backing of $1 million, as Frank and Harry found out when the stockbroker started to ask their Hollywood celebrity friends for money.

Frank and Harry wasted time, jeopardized contacts in the business and destroyed the confidence of a major distributor of syndicated television shows when the whole ship of fantasies sank. They also ended up losing what accounts they had and severing their partnership because they had been players in a fool's game.

They did not know their limitations. They did not know how to tell a legitimate deal from a pipe dream. They strayed outside the bounds of their goals and their real abilities. Neither man had any experience in producing major television movies.

In addition to exceeding their limitations, they forgot one other basic rule of business—never do any work without a contract and a down payment. Also, when someone tells you that $1 million is in the bank for a major project, check to make sure it's there. The producer in the next story did, but he discovered a far more subtle limitation.

## Case 2

The producer's name was Bob and his company was in a small Midwestern town.

He had gone to the Midwest from a successful career on the West Coast to escape the pressures that had given him a bleeding ulcer.

Some of his reputation had preceded him into his tranquil hideaway. Because of this, his business had been relatively easy to launch and to keep afloat at less than a blistering pace.

Before he knew it, the slower pace of life and a 100-year-old farmhouse had hooked him. Five—and then 10—years went by. He lost the ulcer along with all desire to return to life in the fast lane. He also lost the keen edge of his business sense. And this would end up costing him dearly.

Bob had built a fine reputation in his community. His fees were reasonable and the quality of his work excellent. He also practiced much community service. He worked with kids at the local college to give them experience in the field and made a good many free public-service announcements for an assortment of nonprofit local charities.

Bob became something of a regional folk hero, which didn't hurt his business. He also received some national attention in a couple of video and film publications for a project that was billed the country's first community film. The project had been paid for by contributions from area businesses and public-spirited residents, much like a community theater production. This fact made the project newsworthy.

Shortly after the film's premiere, Bob was contacted by a young man we'll call Dan. Dan had a script and a contact in New York. Jack, the New York contact, had $250,000 to invest in making a low-budget, made-for-TV feature. He and Dan figured that the project could be done if Bob would lend his expertise and influence in the community.

Bob was a little suspicious at first. Unlike Frank and Harry from Case 1, he had no ambitions to make it in Hollywood. He had also learned to recognize and avoid pipe dreamers. He agreed, however, to meet Dan and Jack and talk it over.

Jack flew out from New York. In three days of meetings, Bob probed for all the right answers and got them. Jack did, indeed, have the money. He had the key personnel—a director of photography and a director. He needed Bob's guidance and his amazing rapport with the people in the town. With Bob's connections built from the community project—which gave him free use of locations and access to experienced young people in the minor roles and crew positions—Jack felt they could make a movie with a million-dollar look for a fourth of the amount. Bob agreed.

Since Jack wanted to produce, Bob would function as executive producer. It was his reputation and, in a very real way, his town. At the outset, Jack agreed that Bob would have the final say in everything. Jack was relatively inexperienced at production and said that he was relying on Bob to be his mentor. "You'll actually produce it for me," he said. "I'll do all the legwork and learn from you."

With a fee for his services agreed upon in advance, and a down payment in cash, Bob gave the go-ahead and the project began in a blaze of excitement and publicity. Bob arranged for donations of all sorts of equipment, up to and including company cars and a grip truck. He received discount rates on hotel accommodations and provided production offices for the New York crew. He obtained locations and free lunches from several cafes. He assembled a talent pool from which the director chose his cast.

He supplied office help and assistants to the producer and director, got a first-rate camera technician and provided an accomplished art director who agreed to work for just a screen credit. These were only a few of his accomplishments. Two weeks into the project, however, things began to go very badly indeed.

The man from the truck rental firm called Bob and said, "This friend of yours from New York is driving me crazy. He's been over here five times griping about the trucks. I've given him two now. What do you guys want for free? This is the last time I'll help you out, pal."

The woman who owned the laundromat used as a location called Bob to ask him when he was going to replace her carpet because the crew had ripped it to pieces and Jack had said that Bob would take care of it.

An elderly man whose house was used as a location wanted to know when Bob was going to repaint his house (as Jack had told him he would), because the crew had painted the front door and entryway sky blue.

Soon the flood of free and cheap help that Bob had provided began arriving in his office. The lines varied, but typically they ran something like, "Bob, Jack may be a friend of yours, but I can't work for him so count me out."

When Bob tried to talk to Jack about the growing tide of complaints, ruination of his good name and abuse of the town, Jack was no longer rational and reasonable, but instead had become a raving lunatic.

Within a few weeks Jack had alienated nearly every business and individual in town. He had fired or driven away all of the talent and technicians whom Bob had lined up. And in the end he skipped town, literally in the dead of night, leaving behind a stack of unpaid hotel and equipment bills, along with thousands of dollars of uninsured property damage for which Bob, to save his name, had to pay. Needless to say, the salary that was promised for Bob's services never materialized.

This solid, professional, experienced producer learned that there is a limitation to how much real control can be exercised over another adult. Once the project was under way and Jack had what he wanted, he simply rode roughshod over Bob, his company and the community.

Again, the lesson is a simple one. If you spend your time, talent and money establishing a base of respect for yourself in your community, keep it to yourself. And if you're a producer, stick to producing your own work. If you do venture into something of the sort we've talked about here, make sure you carry a big stick and get all your money up front!

You may be the most energetic person in the world. But, as physicists are learning, there is only so much energy in the universe and it does seem to be decreasing. Your energy wasted on pipe dreamers or on unproductive ideas is lost forever. Expend your energy wisely.

The following sections give some concrete suggestions for you to follow after you have made the decision to take this often frightening, but ultimately exhilarating step out of the nest and into the rarefied air of the independent businessperson.

## GETTING STARTED

The first thing you need to do is choose a location for your production company. This decision involves some investigation of the major factors that will weigh on your chances for success.

### The Marketplace

If industrial and sales promotion productions are your specialty, do not choose to set up shop in Two Guns, AZ! The U.S. Chamber of Commerce can help you research the location of large companies throughout the country. And *The Encyclopedia of Associations* gives the location of associations that serve all major businesses in the country; this can be very helpful in your planning.

You don't have to reside in Minneapolis to land a production contract for 3M. But you should know that several production companies are in Minneapolis and they have a better chance of landing an account with this company than you do working from Eureka, CA.

Many people go into business for themselves hoping to be able to live somewhere other than Los Angeles, New York or Chicago. Part of the beauty of the recent trend toward decentralizing America is that many producers are able to operate successfully from smaller communities, where both the cost of living and the cost of doing business are much less than they are in metropolitan areas. None of them can last long, however, if they have chosen sylvan tranquility at the expense of being in close proximity to rather large-scale manufacturing and commerce.

It is possible, for example, for a video production house to be located in the placid, quaintly Victorian Wisconsin town of Oshkosh. Milwaukee, a major marketplace, is only 80 miles to the south and is accessible by both air and a major highway. The owner of the company also has sales opportunities in the corridor of commerce between Milwaukee and Green Bay, which houses major companies such as Kimberly-Clark, Procter and Gamble, Mercury Marine, several major paper companies and a host of smaller manufacturers and merchants.

Assessing the real potential of an area is foremost in deciding where to locate. Once you've evaluated such factors as the number and nature of possible clients you will have to draw on in the area, you're ready to move on to the other considerations necessary to choosing a place.

## The Competition

Some say competition is the spice of life. It is also a very real factor in independent production. Before you decide to jump into the ball game in Hollywood, for example, you should know that the Hollywood Yellow Pages list over 150 independent video producers starting with Ampersand Productions, Inc. and ending with Yukon Pictures, Inc.

In the greater Los Angeles area, the list of production companies runs to 20 pages, 90 companies to the page, from ABA Corp. to Zoetrope Studios. Zoetrope, as you may know, was the brainchild of famous director Francis Ford Coppolla. The fact that it is no longer with us, in spite of its listing in the directory, says another thing about competition. Assessing the competition also means understanding your own limitations.

Assuming that you will eschew the Jacuzzi, the fast lane and Frederick's of Hollywood because you recognize that the market is saturated with competition, there are some things you must do in any geographical area in which you choose to set up shop.

### Evaluate Your Competitors

The first thing to do is to go through the phone book and find out how many other video producers there are in your immediate area and in any major metropolitan center nearby. Check for the numer of local television stations also, because most rely rather heavily on income generated from producing spot commercials and therefore must be considered competition.

You need to know how the business volume of your competition stacks up year to year. Is it steady, increasing or declining? Find out who the primary clients are and if the producers rely on one or two major bread-and-butter accounts or if they have a large number of smaller ones. Check on the size of their operations and sales forces. Do they have many salespeople out beating the bushes day after day or are the owners/producers the head salespeople? Check with the Chamber of Commerce, the

Better Business Bureau and the Franchise Tax Board in your state to learn about the reputations and sales records of the competition.

Visit the site of each competitor. You can tell quite a bit about a company by the way it looks. Does the company have an office, a receptionist and a studio, or does the owner have an answering machine in the den of the house and a magnetic plastic sign on the side of an old Ford van?

What are the strengths and weaknesses of the competition based on your observations? Make a checklist of these and then run through your own capabilities in comparison. If one company has a large staff, an enormous facility and, therefore, a whopping monthly overhead, you might position yourself with advertising that states you can offer substantially lower fees for your services while delivering equal quality because your overhead is lower.

You need to know the present size of the market being serviced and estimate whether it will grow or decline so you can make a sensible guess at what percent of that market you can take as your own.

## Gathering the Information

Gathering all this information about your competition is a major research project, but there are a variety of ways to go about it. It is unlikely that owners will tell you the truth—if they tell you anything—about their operations if they know that you are going into competition with them. The best place to start your research, then, is in a community where you are not planning to locate. If you're honest and not overly pushy, most producers will be happy to give you tips about the business they do since they don't have to worry about you trying to take it away from them.

For the specifics about the competitors in your area, firsthand observation coupled with a little deviousness on your part can reward you with a great deal of factual data. You can, for example, call up a company and say you are a prospective client so you receive a sales pitch. This will certainly supply you with data about the operation and the pricing structure. After you've become a success, however, you may never be annoyed if the same thing happens to you!

If you feel brave and think that honesty is always the best policy, you can even go to a company and say exactly what you're doing. You may receive all the answers you need or you may get silence with this method. Individual owners will react to this approach in different ways.

## Positioning Yourself

Whatever your method, researching your competition is an essential ingredient in the recipe for success. Only by knowing what others are doing, how they are doing it and what they are charging for it can you determine how to position yourself and your service in your market area. In your own mind, as well as in the minds of your prospective clients, try to place yourself somewhere on the scale in between the very best and the very worst producers.

## Where to Set Up Shop

After you have researched the competition, your next step is to choose a physical facility—the options are many.

Stephen J. Cannell has an enormous modern glass tower on a Hollywood Blvd. corner. His name in giant blue letters crowns this edifice and brags openly about his independent production success.

Video Trend Associates, Oshkosh, WI, is in an old Victorian house decorated in contrast to the high-tech nature of video production. From the outside, one would never guess that it was a company at all.

Tech Services, Inc. is a medium-sized production company located in a redecorated turn-of-the-century hardware store in Oshkosh, WI.

Metavision, a very successful producer of industrial programs and music videos, is on the top floor of a K-Mart store on 3rd St. in Los Angeles. Its window is hidden behind the letters "a-r-t" of the glowing department store sign and the owners joke that they are the "art behind K-Mart."

Some one-person operations actually operate out of a spare room or a redone garage at home. And many such producers make a living.

### Determining Your Needs

Your choice of a building should take into account several factors. To begin with, your facility will make an initial impression on clients. If you anticipate clients visiting your place of business frequently, then the shop has to be an advertisement of sorts. You will spend a good deal more on rent and decorations for such an operation than if you decide to take the product to the client and keep your facility just for production and office work.

A production company is not a retail business dependent on foot traffic volume, lots of parking and easy access from public transportation, and therefore the decision of where to locate is much easier than if you were opening a shoe store or a camera shop. Since your overhead—monthly rent, insurance premiums, utilities, etc.—is a primary factor in whether or not you can expect to make it (as we will discuss further under capitalization), you can look off the beaten track for an adequate office.

While you do not want to locate in a run-down area for quite obvious reasons, you might well find an old house for rent in a suitably zoned part of town. If so, you can conduct a small-volume operation with reasonable certainty that you won't be disturbing residential neighbors, and frequently you can find amazing bargains on long-term leasing of older houses. In many cities, downtown businesses are suffering because of the trend toward large shopping malls, and therefore you may be able to find good leasing bargains downtown.

Spend some time driving around town, taking note of promising locations. Read the classified ads in your local papers and call on several realtors. Most important, be sure to check with the local zoning commission to ensure that you will be allowed to operate a service business from the place you choose.

The amount of square footage you will need is entirely dependent upon the nature of your operation. If you think your business will require studio space, for example, then you have to consider ceiling height to enable the installation of a lighting grid. The electrical wiring, if you anticipate any studio lighting, must be much greater than most commercial or residential properties are equipped to handle. The cost of rewiring will be a major overhead and operating budget factor. If you think your operation is going to require a studio, then you should investigate industrial buildings such as former factories, etc.

## Practical Considerations

If your company is going to do most of its production on location, using rental studios when necessary, then your building decision is simplified. Many independent producers have no permanent full-time staff; they simply hire free-lance technical help on a per-project basis. Others keep a regular staff in-house and on salary. Your operation will probably fall between the extremes possible on both ends of the scale. No matter where your operation fits in, here is a checklist to help you determine your physical requirements.

- *How much square footage will you need?* At a minimum you should have enough space for a private office for yourself, a reception area, private working areas for each of the employees you anticipate hiring (or using frequently), a screening/conference room and storage space for equipment, supplies, videotapes, etc. Bear in mind that much of today's high-tech equipment, as well as those precious tapes, require air conditioning and dehumidifying if you want to keep them in top shape.
- *What kind of parking is available?* You should make sure that the location has a parking lot or plenty of free and unlimited parking on the street for you, your employees and clients. If your operation is going to have a van, a grip truck or any other specialized vehicles, then you will also need to consider garage facilities for them.
- *What about remodeling?* It is unlikely that you will walk into a former store or old house and find it suited to production needs. When considering a lease arrangement, find out if the landlord is willing to remodel to suit your needs at no cost to you. The remodeling may become a bargaining chip for reducing your

monthly rent if you are handy and can do the work yourself. Don't be afraid to try a little old-fashioned horse trading or haggling here. Remember that you are going to be spending a considerable portion of your income on the physical facility and you have the right and the obligation as a businessperson to try to get the best deal possible.

• *Why is this the right location for you?* This is the main question to answer for yourself. The worst thing you can do is to make do, thinking that you will move into the right place later on. A business has to project an image of stability, and staying at one location for a good deal of time helps foster that image. Moreover, once you are in business, it is tremendously expensive to pick up the staff, the office furniture, the equipment and so forth to make a move across town.

Take some time making your location decision. Don't rush into it in the excitement of getting started. A few extra days or weeks of waiting and doing some soul-searching and property searching can make your move a good one. Remember that you are going to be spending much more time at the business than you are at home, so you want the place to make you—as well as your employees and clients—feel good about being there.

## Finding Equipment Suppliers

The next step, now that you've assessed the marketplace and the competition and found a physical facility to suit your purpose, is to take stock of your suppliers. Every business needs to know what to get and how to get it at the best possible price. As an independent video producer, your requirements go beyond those of a retail business, which has to consider only stock, office supplies, etc.

In your business you must also consider a dizzying array of equipment coming out each year that is doing three things simultaneously: getting better in quality, lower in price and making what preceded it the year before obsolete. Keeping up with technological innovations and suppliers in your field is one of your tasks.[3]

You can fill file cabinets with the literature and brochures that manufacturers and distributors are generally happy to send you on request. Even if you are not really in the market for anything, it is a vital part of your job as an independent producer to know what the technological innovators are doing to provide continually new and improved tools for your trade.

Cameras, for example, are becoming more and more miniaturized and less and less expensive. It is perfectly feasible for any producer on a very modest budget—for a total outlay of $12,000—to own a high-quality, three-tube or three-chip video camera, a videotape recorder, a simple-to-use, direct-cuts-only editor and a very satisfactory audio board with mixing and equalization capability.

## Renting Equipment

Even if you decide to own some production equipment, it is unlikely that you will go much beyond a few basic tools. It makes no sense for a producer who does only three or four major productions a year, filling in the remainder of the time with less complex and smaller projects, to own lots of expensive equipment that requires maintenance and which, when it is not being used in the field, represents a net loss.

Throughout the country there are rental houses that have everything you need from a second camera to a complete mobile production studio (as discussed in Chapter 1). Your production library should include the *current* catalog from each rental house, since prices and availability usually change annually.

### Credit Accounts

You will want to establish a credit account with the primary rental house you choose, just as you will with all of your other suppliers. Most business credit accounts are payable 30 days after the billing date. In some cases you will find 60- and 90-day payment schedules that allow you to collect on your accounts receivable before you have to pay your accounts due. Even if you have collected your money, it's wise to leave it in an interest-bearing account for as long as you can before paying it out to someone else!

Credit in business is absolutely essential to its success. Since video production equipment is so expensive, and because this business has its share of fly-by-night operations that abuse the privilege of credit, it is not easy to get a credit account with a rental house. You will be required in most cases to provide the following:

- Your personal credit history
- Your business credit history with three references from banks or other suppliers with whom you have accounts (this can be a Catch-22 if you are just starting out)
- Proof of insurance for the maximum value of all the equipment you intend to rent
- A substantial cash deposit if you have no business credit history

Once you establish a credit account with your rental house, maintain it as arranged (of course this axiom applies across the board to all of your suppliers). If your name ever pops up in a computer file because you skipped a bill, your future in this business may be in jeopardy. The world of production is quite small, in spite of the numbers of people who are in it.

### Expectations

If you are paying your bills on time, you can also expect and demand to have the equipment you rent be in top-notch condition and to be available on time, every time. Have your technicians check each piece as it is unpacked to be sure that it operates as it should. If it doesn't, or if it didn't arrive when promised, call immediately and let the rental house know. Every good rental house wants you to be totally satisfied. If yours doesn't, choose another one.

A final note about renting is that in most cities you will also find office furnishings and machines available for rent or lease rather than purchase. Exercising this option can mean a major reduction in the amount of capital you will need up front to begin a business.

## Locating Technicians and Talent

### Technicians

Briefly put, you can't make a video on your own. You need a stable source of people—videographers, assistant directors, editors, recordists, gaffers, etc. A final consideration when deciding on your location is the availability of free-lance technical talent.

There are regional business offices for the trade unions and guilds we described in Chapter 1. If you are a union contract signatory, help for any project is only a phone call away. In Appendix C, we review most of the major publications in which you can find Situation Wanted ads, or can place your own Help Wanted ads to develop a reference file of personnel.

It is important that the place you choose to locate is close enough to the labor pool to make it profitable for you to pay transportation costs for each free-lancer you will need. If the nearest reliable pool is several hundred miles away from your location, but you believe that your choice is suitable in every other respect, then you should consider hiring a small, full-time technical staff.

Devise a current and projected personnel plan to help determine your labor needs. Consider what skills you have, what you will need, whether your employees will be salaried or hourly and what the legal and financial ramifications are going to be in terms of withholding taxes, fringe benefits, etc.

### Talent

As for talent, if you only want to do documentaries or industrial shoots featuring the actual people who work at the Land O' Lakes cheese factory, for example, talent is not going to be a problem. If you look forward to doing television spot commercials or any other kind of production (like the Jones Hotel training video with actors), then you need to know where to find talent.

As with technicians, if you are a union contract signatory then your problems can be solved by contacting SAG, AFTRA or the American Federation of Musicians at one of their regional business offices.

For nonunion productions you may still be able to find some talent agencies that book nonunion members. Since talent agencies are really employment agencies, in most states they require licensing. Check the appropriate bureau in your state for its requirements. Since licenses are a matter of public record, you will be able to find a list of licensed talent agencies.

Another option is to check with local theater groups and college drama departments; you will be in for a happy surprise. The fact that these performers are not in Hollywood does not mean they are without talent.

Placing ads in some of the publications discussed in Appendix C will also help you build a file of available talent. Your ad might read something like this:

> Video production company seeking talent pool for commercials and other anticipated productions. Send composite photos and resume to: My Own Production Company, address and city.

### Free-lance or Permanent Staff?

There are pros and cons to using both free-lance technical talent and permanent staffers. Here are some final considerations you might want to mull over before you decide which way to go.

Some people feel that free-lancers are not as professionally competent as those technicians who work full time for a company. In general, this is not the case. Many free-lancers are, in fact, seasoned veterans who have become tired of the rat race and who prefer to operate independently, picking and choosing their jobs and their schedules.

Sam Drummy, for example, is a two-time Emmy Award winning videographer. After working full time for NBC for a number of years, he chucked the security of corporate life to free-lance. In his mid-forties, Sam works as many days a year as he likes, makes more money than he did with the network and splits his time between his home in the Hollywood Hills and the island of Hawaii.

Although Sam Drummy's credentials are above average, there are thousands of highly skilled craftspeople who free-lance in our industry. A good way to check the qualifications of a free-lancer is to look at the person's list of credits and the ubiquitous demo reel that any real pro will be happy to show you.

One major advantage of hiring free-lancers is the relief from much of the paperwork and financial burden of full-time employees. A free-lancer is an independent contractor. You do not need to withhold state or federal income tax, pay for fringe benefits, unemployment insurance, etc., (except in the case of union members, for whom you must pay into the guild or union fund). Only if you pay one person more than $600 in a single year are you required to file a form 1090 with the IRS (see your accountant). Another advantage to having a free-lance staff comes during slack business times when you are not obligated to pay someone a salary to sit around the shop doing little, if anything.

A major disadvantage of using free-lancers is that they are not always available when you want them. If you are scheduling a shoot, you may have to do so around the free-lancer's previous bookings. If a project springs to life suddenly and you're used to working with a certain videographer who happens to be in Borneo that month, you'll have to settle for a different person.

In order to avoid delaying a shoot while you wait for the return of your itinerant free-lancer, make a file from which you can draw several good people in each category. Sample the demo reels or other material from a large number of people and keep data on the best prospects in your computer or Rolodex. It is unlikely that all will be on assignment at the same time.

Full-time staffers are beneficial for a number of reasons. To begin with, it is easier to develop a family atmosphere with a permanent staff. If people feel that they are part of the group, working toward shared, long-range goals, you can expect them to work longer hours, sometimes not necessarily for pay. Since they are part of your team, involved in the day-to-day operations and expectations of the company, you can also expect to lead them to a consistent modus operandi and point of view—perferably your own.

Planning becomes easier when you have a core of dependable people on hand at all times. Since everyone is working for the good of the company and not just for the daily rate, you can share ideas and tasks. By delegating tasks and sharing ideas, you can accomplish more.

Permanent staffers are more likely to make contributions of time and talent and ideas to promote the common cause, provided that you make them integral parts of the company, and not mere functionaries working for paychecks. An example of this attitude comes from one of the permanent staffers at Aaron Spelling Productions in Hollywood. "The only way someone's going to get a job here," he said, "is for one of us to die or retire. Aaron doesn't have a company, he has a family. And we're all part of it." Certainly, this type of employee attitude contributes to the continuing phenomenal success of this independent producer.

In the final analysis, as with most producers, you will probably end up using a mix of free-lance and permanent help. The beginning producer simply can't afford to take on a large staff. At the outset, you will use free-lancers almost exclusively. As your company grows, take on permanent staff in the most vital areas. Your profit and loss sheet will dictate, to some extent, just how large your permanent core will become. Affordability is not the only criteria to use, however. Consider one final story.

Theo Mayer and Peter Inebnit at Metavision experienced a very rapid surge of growth early in their history. From their humble beginnings (with $500 and an occasional free-lancer), they soon found their company with 24 full-time employees.

"While we were making tons of money," Mayer said, "we were also working 18- to 24-hour days, seven days a week, and spending it all just to support the tremendous overhead we found ourselves with."[4]

Inebnit agreed. "Somewhere along the way we found that we'd forgotten why we wanted to be in business in the first place. We were doing jobs we didn't like, Theo was out pounding the streets every day bringing in more, and one day we just sat down and looked at each other and asked why we were doing this."[5]

"We went into business," Mayer concluded, "to do good, creative things, to have fun and to take care of ourselves, our families and our partners. We never dreamed of being responsible for 20 or 30 other people and their families along the line."[6]

Mayer and Inebnit, caught in the paradox of too much success, fired most of the 24 permanent staff of Metavision. They sold the bulk of equipment they had purchased to support the large operation and returned to their original goals, taking on only as much work as they could handle personally, with free-lance help as needed. Their story is a good one to remember.

## Regulations and Licenses

The regulation of small business varies from city to city, state to state and business to business. In general, though, you can anticipate needing a business license from your city, town or county.

At the state level you will probably need to register under a fictitious name or "doing business as" law unless your full name appears in the title of the business. Most states will also require you to file for a sales and use tax number.

Finally, contact the Internal Revenue Service (IRS) for a federal employer's identification number and a "Going Into Business Tax Kit." The latter is a free service from your friendly tax collector.

In some places around the country you will find the business climate to be very hospitable, with a minimum of red tape for you to cut through. In others you will find the restrictions so frustrating that you may have to add psychotherapy bills to your overhead. In spite of what you may hear about "deregulation," a certain amount of regulation still exists and governing bodies must be dealt with at the outset of your business venture.

While there is no rule that you need a lawyer to go into business, it is very prudent to talk to one. Interview several lawyers to find one with whom you are compatible, because your relationship with him or her will probably be a long one. A lawyer can take care of many headaches, not only as you launch your business, but as you operate from day to day and year to year.

## Capitalizing the Business

A business should be able to keep its doors open for a minimum of one to two years from the date it starts without making one dime of profit. Undercapitalization is the number two reason for business failure, only a fraction behind poor management!

Acquiring capital and accounting for it are the final phases before opening the doors of your new business. All of the things you have done up to now will bear on the success (or failure) of your capitalization efforts.

Use the information you have compiled thus far to develop a business plan. Figure 3.1 is a worksheet from the SBA's *Checklist for Going Into Business*, which will help you determine the amount of money you will need to start.

### Choosing the Right Legal Setup

The right type of capitalization depends upon the type of business setup you choose. The three legal forms are: sole proprietorship, partnership and corporation.

The form of business you pick depends on such things as your personal financial status, the number of employees you anticipate, the risk involved and your tax situation. Consult your lawyer and accountant for advice on which form is best for you.

A sole proprietorship means that the full burden of capitalization and all of the risk are yours. A partnership can help you share these, but unless it is a limited partnership, you will give up sole control of your destiny. Standard corporate structure may make you accountable to a board of directors and can be frightfully restricting to your freedom, especially if you issue stock to the public.

There is now a setup called the Subchapter S corporation designed for small businesses. This type of incorporation treats profits or losses by the corporation as ordinary income or loss to the individual stockholder. You find investors who need tax write-offs during a period when you are most likely to suffer losses. As you gain financial ground you begin to buy back the stock. At any time, you are free to change to a regular corporate structure. One of the beauties of this arrangement is that in order for investors to keep their tax advantages, they must let you direct all the affairs of the business as you see fit.

## Figure 3.1: Worksheet for Estimating Capital Needed to Start a Business

# Worksheet No. 2

| Estimated Monthly Expenses<br><br>Item | Your estimate of monthly expenses based on sales of<br>$ _____<br>per year | Your estimate of how much cash you need to start your business<br>(See column 3.) | What to put in column 2<br>(These figures are typical for one kind of business. You will have to decide how many months to allow for in your business.) |
|---|---|---|---|
| | Column 1 | Column 2 | Column 3 |
| Salary of owner-manager | $ | $ | 2 times column 1 |
| All other salaries and wages | | | 3 times column 1 |
| Rent | | | 3 times column 1 |
| Advertising | | | 3 times column 1 |
| Delivery expense | | | 3 times column 1 |
| Supplies | | | 3 times column 1 |
| Telephone and telegraph | | | 3 times column 1 |
| Other utilities | | | 3 times column 1 |
| Insurance | | | Payment required by insurance company |
| Taxes, including Social Security | | | 4 times column 1 |
| Interest | | | 3 times column 1 |
| Maintenance | | | 3 times column 1 |
| Legal and other professional fees | | | 3 times column 1 |
| Miscellaneous | | | 3 times column 1 |
| **Starting Costs You Have to Pay Only Once** | | | Leave column 2 blank |
| Fixtures and equipment | | | Fill in worksheet 3 and put the total here |
| Decorating and remodeling | | | Talk it over with a contractor |
| Installation of fixtures and equipment | | | Talk to suppliers from who you buy these |
| Starting inventory | | | Suppliers will probably help you estimate this |
| Deposits with public utilities | | | Find out from utilities companies |
| Legal and other professional fees | | | Lawyer, accountant, and so on |
| Licenses and permits | | | Find out from city offices what you have to have |
| Advertising and promotion for opening | | | Estimate what you'll use |
| Accounts receivable | | | What you neeed to buy more stock until credit customers pay |
| Cash | | | For unexpected expenses or losses, special purchases, etc. |
| Other | | | Make a separate list and enter total |
| **Total Estimated Cash You Need To Start** | | $ | Add up all the numbers in column 2 |

Courtesy of the U.S. Small Business Administration.

Internal Revenue Service Code Section 1244 allows you in this situation to treat losses on the stock of a "small business corporation" as deductions against ordinary income. IRS publications 542, 544 and 550 have sections discussing this regulation. Since the rules for taking advantage of Subchapter S and 1244 are involved and very specific, read all the publications and talk them over with  an attorney before you decide.

## Possible Funding Sources

Aside from incorporation with stock offerings, there are four major sources of funding your business. You (or you and a partner) can (1) put up the money yourselves, or you can secure loans from (2) a bank, (3) the SBA and/or (4) the Veterans Administration (VA).

According to University of Wisconsin business professor Richard Krueger, financing through a local bank is the most likely to succeed. Banks like to see local business starts because they help keep money in the area. This is especially true if your business enhances or stimulates other businesses in your area.

The SBA has money to loan, but the amount available is limited and is highly affected by political factors from one administrative term to another.

While the Veterans Administration has money to lend to qualified vets for business reasons, there are two myths about VA loans. The first myth is that loans are plentiful; the second is that they are easy to get. In addition, VA loans have some requirements that are often difficult to meet, such as the stipulation that you live on the site of your business. If you can operate out of a basement, spare room or garage, investigate the VA opportunities in your area.[8]

There are other funding sources often overlooked by producers. For certain projects you can get money from private foundations, as well as from state and local governments. The following publications catalog most of the major grants that might apply to you and your production company: *The Grants Register* (St. Martin's Press, 175 5th Ave., New York, NY 10010); *The Foundation Directory* and *The Foundation Grants Index* (The Foundation Center, 888 7th Ave., New York, NY 10106); and *The National Endowment for the Humanities* (1100 Pennsylvania Ave. NW, Washington, DC 20506).

Check with your state government for information on grants and endowments that it may administer. Nationwide, millions of dollars are given out annually for media projects, many of which you have seen on Public Television. There is no reason why you can't attract some of this money to fund worthwhile projects of your own. Writing proposals for grants is an art; many colleges offer classes in this specialty. Some granting institutions will give you copies of previously successful grant applications to use as a model for yours.

## Preparing the Presentation Package

Regardless of which money source you choose, make a presentation to the prospective lender. The following outline shows the format followed for a typical SBA guaranteed bank loan:

I. Summary
   A. Nature of the business. In this section write a statement summarizing what your research of the market has shown. The full report will be in your business plan.
   B. Amount and purpose of loan. Using the SBA worksheet in Figure 3.1, estimate how much it will cost to set up and operate for a year.
   C. Repayment terms. Estimate a reasonable time and amount for repayment that will not cripple your operation.
   D. Equity share of borrower (debt/equity ratio after loan). It is extremely unlikely that any lending institution will grant the entire amount. You will be expected to have up to one-half the total as your show of good faith. Equity in your business does not have to be cash. It can include any equipment or facilities that you have on hand. Signed contracts or letters of intent from clients and/or distributors may also be bankable equity.
   E. Security or collateral. List here all of the equipment, including office furnishings, etc., that you intend to purchase with the loan. Get market value estimates and quotes on these costs. You may also be required to put up some of your personal property in some cases, such as a second mortgage on your home. (Be very sure of your chances for success before going that far!)
II. Personal information
   This information is provided for all corporate officers, directors and any individuals owning 20% or more of the business.
   A. Education, work history and business experience.
   B. Credit references. This will include your personal references (credit cards, automobile financing, etc.) as well as any business credit you have established.
   C. Income tax statements for the past three years.
   D. Personal financial statement. Your statement must not be more than 60 days old.[9]
III. Firm information (for whichever is applicable A, B or C).
   A. New Business
      1. Business plan.
      2. Life and casualty insurance coverage.
      3. Lease or purchase agreement on facility.
      4. Partnership, corporation or franchise papers if applicable.
   B. Business acquisition (buyout of existing firm).
      1. Information on acquisition.
         (a) Business history, including seller's name and reasons for sale.
         (b) Current balance sheet.
         (c) Current profit and loss statements.
         (d) Business federal income tax returns for the past three to five years.
         (e) Cash flow statements for previous year.
         (f) Copy of sales agreement with breakdown of inventory, fixtures, equipment, licenses, good will and other costs.
         (g) Description and dates of permits already acquired.
         (h) Lease or purchase agreement on facility.

2. Business plan.
3. Life and casualty insurance.
4. Partnership, corporation or franchise papers if applicable.
C. Existing business expansion.
  1. Information on existing business.
    (a) Business history.
    (b) Current balance sheet.
    (c) Current profit and loss statements.
    (d) Cash flow statements for previous year.
    (e) Federal income tax statements for last three to five years.
    (f) Lease agreement and permit data.
  2. Business plan.
  3. Life and casualty insurance.
  4. Partnership, corporation or franchise papers if applicable.
IV. Projections
  A. Profit and loss projections (monthly, for one year) and explanation of projections.
  B. Cash flow projections (monthly, for one year) and explanation of projections.
  C. Projected balance sheet (for one year after loan) and explanation of projections.

### The Five Cs

After you have prepared your loan package, take a critical look at yourself to see if you make up the typical "good risk" candidate for a major loan. The "Five Cs" of this person are:

1. *Character:* If you have a history of being honest, have a good credit rating, are experienced in the field of video production, are ambitious and communicate well, you are a person of good character.
2. *Capital and collateral:* Your financial statement is your guide here.
3. *Capacity:* Assess your expectations and the funding source you expect to use to repay the loan in a timely manner.
4. *Conditions:* The conditions here are your health, and the marketability of your products and services versus those of the competition.
5. *Consequences:* Will the loan be truly productive? Can you demonstrate to the lender that both you and it will benefit in the long and short term?

If your self-assessment is positive, you have an excellent chance of convincing your local bank to help make your production company a reality. Keep your expectations conservative. Any debt in business, just as in personal life, should be acquired cautiously. Too much can cripple you. Too little may restrict your ability to function.

## Accounting and Record Keeping

Accounting and record keeping are often the weak links in an otherwise good operation. These two vital elements organize and control your business. It is essential to keep good records for these reasons:

- To let you know if the business is making or losing money
- To enable you to make solid management decisions
- To obtain more financing
- To comply with local, state and federal laws and regulations
- To make your case if you wish to sell the operation

Since this area of business operation has become so specialized and endures a constant stream of changing rules and legislation (especially in the tax area), the best thing to do is to put yourself in the hands of a competent accounting firm. Your accountant will provide you with a bookkeeping system or will provide regular monthly service depending on your need.

The accountant will determine the best method for depreciating your fixed assets, show you how to keep every penny allowable by law from the tax collector and worry about many of the details that would otherwise get in the way of your creativity.

At a minimum, your records will include cash receipts, cash disbursements, sales, purchases, payroll, equipment owned, equipment rented, inventory, accounts receivable, accounts payable, sales tax and withholding tax for you and your employees.

You may be required to make quarterly tax statements in April, July, October and January. You may also have to make a major payment to the IRS based on your estimated profit for the year if you are not on a tax withholding system.[10]

Once you have a good accountant, your role in accounting and records keeping will require little more than reconciling your bank statement monthly, depositing all income in a timely manner and getting on with your work.

Be certain that you keep your personal records and accounts separate from those of the business. Failure to do so is an invitation to trouble. Remember that the business is not you; your salary—along with all the other expenses—come out of it. That's why you program a profit margin for the company into every budget.

In 1974 the Commission on Federal Paperwork reported that small business in America was paying between $15 and $20 billion a year to fill out reports and forms. In the years since then, business has been somewhat relieved. Reporting remains, however, a major annoyance factor as well as a costly one.

Find out what forms you will be required to file and set up a calendar of report due-dates so you know when they have to be completed. The bigger your operation, the more forms will be required. Table 3.1 lists the forms that sole proprietors must usually complete for the federal government. Check with your state for its requirements.

**Table 3.1: Tax Forms Schedule for the Federal Government**

| Form | Number | Date Due |
|------|--------|----------|
| U.S. Individual Income Tax Return | 1040 and Schedule C | April 15 |
| Declaration of Estimated Tax for Individuals | 1040ES | April 15 and quarterly |
| Employer's Annual Federal Unemployment Taxes | 940 | January 31 for preceding year but payments may be due quarterly |
| Employer's Tax Return | 941 | Quarterly with payments likely due monthly |
| Wage and Tax Statement | W2 | To employee by January 31 for preceding year |

## SETTING AND REACHING GOALS

As you can see, setting up shop is not a simple task—neither is goal-setting. Goal-setting is not merely an academic formality; setting *and* reaching goals are two of the most important elements in setting up shop.

The independent producer has to assess real, rather than wished-for (or imagined) capabilities. How ambitious can you really expect to be, given your capitalization, equipment, location, available talent, initiative and other resources? This is all part of discovering your limitations. Only you can determine them, and doing so takes a good deal of honesty. When you know your limitations, you can establish a system for setting realistic goals and then striving to achieve them according to a regular schedule.

In your business plan you have projected some of your goals. Included in these goals are an estimate of the amount of business you hope to do, what costs you anticipate incurring, how much income you see as a result of the successful implementation of the plan, etc. Each month, evaluate your status in relation to that business plan.

Let's say you expected to have four accounts in hand by the end of the first quarter and you have only two. You obviously have a problem meeting the goal you set for yourself. Analyze the situation and determine the reasons. Perhaps you've spent too much time on one account at the expense of the others you had in mind. Maybe you've encountered sales resistance from potential clients. Get back to them and find out what you can do to overcome it.

Check the records of your time spent during the period in question. Ask yourself how much of it was devoted to the active pursuit of those clients. If you find that much of the time was wasted on hopeless causes, putting things off until tomorrow, dealing with problems that could be delegated to someone else in the organization, etc., work on using your time more productively. There's nothing hard or mysterious about this angle of the business. All it takes is the ability to assess accurately what you should be doing and then doing it.

Make a list of personal goals aside from those stated in the business plan. It may sound silly, but writing goals on paper helps to make them concrete. Here are some sample goals other producers have set for themselves:

- I want to feel good about myself and my business
- I want financial independence in five years
- I want one month off every year to go fishing
- I want to manage my time efficiently
- I want three days off every week by the third year of this business
- I want to learn to handle stress better
- I want enough time to enjoy my family
- I want my employees to respect and like me
- I want people to know that my word is my bond
- I want to do award-winning work

Notice that each item begins with, "I want." Once you've determined your own goals, you have a basis for implementing plans to achieve them. Don't expect them to remain the same year after year, though. Establishing goals should be an ongoing process. Writing them down regularly helps you keep sight of the real ones. It also helps you recognize those goals that become less important as you continue to grow.

There is one danger in setting and keeping goals for both yourself and your business; you risk losing your flexibility. Resist this pitfall with all your might. Keep your mind open to suggestions from your employees, friends, business associates and your family, even if they don't fit your preconceived goal statement. Many opportunities have been missed by people who get so wrapped up in achieving their goals that they can't see that a new one presented to them might just be better.

The function of goal-setting is not to cast you or your company in cement. It is to give you a touchstone on which to affirm the major reasons why you are an independent producer and a yardstick by which to measure your progress.

## SUMMARY

As you can see, setting up shop is not a simple task. In this chapter we have provided a method to help simplify it for you. No one can guarantee that you will make it in business, of course; but others have blazed a trail before you and there is help available.

As you begin the process, remember that the U.S. Government is really on your side. Call on the Small Business Administration for free information and advice.

Finally, don't be afraid to go back to school. Many universities and community colleges offer night courses in a wide variety of business subjects. You'll find them informative, frequently fun, and a good way to meet other people who may be going through this process with you.

Being in business for youself will require all your skill, organizational ability and belief in yourself. It is frightening at the outset, but, if you dare to venture forth, the rewards for having done it on your own may bring you the most satisfaction you have ever known.

## NOTES

1. U.S. Small Business Administration, Fort Worth, TX, 1984.

2. *Checklist for Going Into Business* (Fort Worth, TX: U.S. Small Business Administration, 1983). For a free copy of this publication, write to the SBA, PO Box 15434, Fort Worth, TX 76119.

3. For annual listings of video equipment suppliers and manufacturers, see *The Source Annual Buyer's Guide* (New York: Broadband Information Services, Inc., 1985) and *The Video Register 1985-86,* by the editors at Knowledge Industry Publications, Inc. (White Plains, NY: Knowledge Industry Publications, Inc., 1985).

4. Theo Mayer, interview with author, January, 1984.

5. Peter Inebnit, interview with author, January, 1984.

6. Theo Mayer, interview with author, January, 1984.

7. For help in formulating a business plan, see the *Checklist for Going Into Business* (Fort Worth, TX: U.S. Small Business Administration, 1983).

8. For further information on government loan sources and requirements, request the booklet *Loan Sources in the Federal Government* (Fort Worth, TX: U.S. Small Business Administration).

9. For assistance in preparing a financial statement, see the *Consumer Information Report #5, How to Prepare a Personal Financial Statement,* available from Bank of America, Consumer Information, Box 37018, San Francisco, CA 94137.

10. For a discussion of the tax ramifications of sole proprietorship, partnership and incorporation, see *Circular E of the Employer's Tax Guide,* available from local IRS offices.

# 4 Promoting and Marketing Your Company

One axiom in business goes like this: "There are those who advertise and those who go out of business." A production company is no exception to this rule. While advertising and promoting will not guarantee your company's success, the absence of advertising and promoting will almost surely cause it to fail.

Unlike other businesses, the independent production company must reach a very specialized clientele. If you were running a shoe store, advertising your product would be very simple. You would plan a yearly strategy employing a media mix that includes newspapers, throw-away shopping papers, radio and TV. You would try to reach all the potential shoe buyers in your market. Reaching all the potential clients for your video production house is not so simple.

Production houses, like ad agencies, remain behind the scenes. They are largely unknown to the public. There aren't any credits on TV commercials to let viewers know who created them. Placing ads in newspapers and magazines with sufficient repetition to reach your target audience is a frightfully expensive proposition that most small producers can ill afford.

There are some conventional advertising tactics that you can use, of course. This chapter will discuss these, as well as a variety of creative methods that will promote your company at little cost to you. Some examples of effective promotional techniques that have been used by other producers will also be provided.

## PRINT ADVERTISING

The sort of print advertising you undertake will be dictated by two primary considerations: your target market and how much money you have to spend. Advertising and promotion costs must be a part of your annual operating budget. While there are no "standard" figures for how much to include in this category, most conventional businesses budget between 5% and 15% of their anticipated gross receipts for advertising. If you are starting out in a highly competitive location, you may have to spend somewhat more than this amount to position yourself in the marketplace. Set a reasonable figure that fits your specific situation.

After calculating a dollar amount to spend, the next step is to figure out where to spend it. This decision will depend on your advertising strategy. That means, very simply, getting the right message to the right buyer at the right time.

In Appendix C, we review a number of publications that receive wide circulation in the industry. Since the readership of these publications consists mainly of producers, most of their advertising is directed toward you, the producer. The ads, for the most part, are for firms that offer services and supplies for production and postproduction. Take a look at each one of these publications. One or two might have the type of circulation for which an announcement of your existence is appropriate.

As ridiculous as it sounds, you have to beware of your ego here. Some producers place ads in slick magazines and periodicals simply because some big-name, big-time companies are in them also. It makes them feel important to see their name alongside that of David Wolper.

Placing an ad in a publication directed at show-biz types, such as *Variety* or *The Hollywood Reporter,* is almost pure egomania. Your friends and family may be impressed when they see your name there, but the potential client whose eye you wish to catch probably will not see it. The same is true if you leap in buying quarter-page spreads in the big, slick national publications geared to the video trade. The businessperson who needs a sales promotional or training tape probably won't see it.

It makes more sense for you to advertise in journals and papers that reach your target market. If, for example, you plan to make your living doing video depositions for trial lawyers, place a small, tasteful ad in the journal of your state's Bar Association.

You can make a good living in many locations doing real estate videos. Advertise for this market in the bulletin or journal of your local and regional real estate association.

The point here is simple. Choose your target, find the publication that serves that target and advertise there.

## Designing Effective Ads

Designing an ad for your company should be simple. You know your capabilities and your limitations. You know what your pricing structure is and how it stacks up against the competition. You have two options in print: the simple classified or the display ad. Since a classified is rather simple to construct, we will concentrate on designing a display ad, which can normally be run in full-, 1/2-, 1/3-, 1/4-, and 1/16- page dimensions.

Many people gifted with the ability to produce film or video have no idea about the fundamentals of print advertising. Larger publications come to your aid by providing copywriting and layout services when you buy an ad from them. Remember that these people are professionals in their area; don't hesitate to turn over the project to them. Smaller publications don't always offer this service. They expect you to give them a camera-ready layout. All they do is shoot it, size it if necessary and print it.[1]

In general, remember two things about making an ad. First, know the personality of the intended reader. Second, understand that your ad is in tremendous competition for the eye of that reader.

If you're trying to reach conservative businesspeople, your ad must have a different feel and texture from one designed to reach creative directors at advertising agencies. With the latter you might want to inject some humor as an indication that you recognize their position. For example, use a picture of you and your staff doing headstands with the caption, "We'll stand on our heads to get your attention!" Things like that work with creative types. The photo would be followed, of course, by copy telling creative directors just why they should pay attention to your company.

With businesspeople, you have to be much more subtle in implying that you are creative. Their concern is how reliable and businesslike you are.

In any display ad, a good-quality, high-contrast photo is almost essential as an eye-grabber. What you choose to show depends again on whom you wish to impress. A dynamic shot of you and your crew on location in a busy factory might work to attract the eye of an industrial client. A tight shot of a hamburger with a video camera behind it might be the thing for one of the agencies handling fast-food accounts.

Try not to generalize. Ads that proclaim "We do everything for everybody" usually receive a "Who are they trying to kid?" response. Evaluate your own feelings about such broad claims. Tailor your ads for each specific market. The hotel owner will feel more comfortable believing that you specialize exclusively in hotel training aids. Make the realtor believe that all you do is real estate, etc.

None of this implies that you can be fraudulent. If you do a wide variety of video work, do not say anything to the contrary in your advertising. It is perfectly all right, however, to talk about only one of the areas of your work in each ad.

## Repetition Counts

Once you have selected the publications that you feel will do the job, stick with them. It takes multiple insertions to reach your reader. Each publication gives you circulation figures so you can calculate your cost per reader (this "reach" is normally calculated in cost per thousand readers). Your cost per reader will be much higher than the cost per thousand for a breakfast cereal maker, obviously, because your target audience is much smaller. The return on a successful ad, however, is a great deal more than the cost of a box of Wheaties!

In order to increase your chances of an ad resulting in a contact and a sale, you must be able to sustain it over a series of repetitions. It will do you no good at all to place an ad once. It must appear over and over again during the course of a year. This gives your company a solid, stable image in the field.

Remember that the main function of any kind of advertising is to place the name of the advertiser in the mind of the buyer. If you want some tomato soup, the first thing that comes to your mind is probably the name Campbell's. Other companies trying to sell tomato soup know that and have to position their soups against this industry leader.

One of the reasons for the Campbell's phenomenon, aside from quality and price, is that the name has been repeated over and over again for a long time in advertising. Every single product, service or manufacturer that pops into your mind when you think of names like Kleenex, Jello, 3-M and Star Kist does so because of repeated advertising.

Bear in mind that advertising alone will not sell the products, nor will it sell your production company. What it will do is establish the name of your firm in minds of potential clients. If and when they do decide they need video production services, you will have a much better chance of getting the call or being accepted when you make inquiries because your name is familiar than someone who considers advertising dollars a waste of money.

In the battle analogy that many businesspeople use to describe the fight for survival, advertising your name is the offshore artillery used to soften up the beachhead you hope to establish. It's as simple as that.

## PROMOTIONAL LITERATURE

There are three primary pieces of printed material that you can use: the business card, company letterhead and brochure. Each is a "first-impression" item because a potential client will draw an impression of you and your company from these things. Because of this, the concept and execution of each piece of promotional material demands your best effort.

At this point you need to think about the image you want to project. This, in large part, is determined by the name you choose for your company.

The natural tendency for many people opening a business of their own is to put their name on it. After all, it's your money, your time, your gamble and your statement of freedom, so why not call it Sally Smith Productions, if your name happens to be Sally Smith? You certainly *can* use your name in your business, but it may not always be the best idea. Stephen J. Cannell and Aaron Spelling call their company Cannell and Spelling, but they were successful writers/producers before they went into business for themselves. They had track records of hit shows. Their names on the front door and the letterhead meant something to prospective clients.

Ron Bullock from Oshkosh, WI, did not have a name that would sell. When he went into business he chose a company name and logo designed to convey first, a sense of what he did, and second, an image of stability.

## Designing a Business Card

The card designed by Bullock is unique and tasteful (see Figure 4.1). It is a conservative card because the Fox River Valley of Wisconsin in which he does the bulk of his business is a traditionally low-key, conservative place.

The card contains the basic information underlined with three horizontal lines giving the card a bottom-heavy solidity with an airy use of negative space above it. The ink is also raised giving the card further dimension and texture. Bullock chose the name, as he says,

> Because I wanted to have the word video in it, because the word "trend" has a positive connotation to a lot of people who want to feel like they're in on something and because the word associates is used in many of the businesses, such as law firms, that I wanted to attract. That word is also pertinent to my operation because I'm essentially a one-man band and I employ creative associates on a per-project basis.[2]

Bullock's use of the job title "production manager" on the card is a clever ploy.

**Figure 4.1: Sample Business Card**

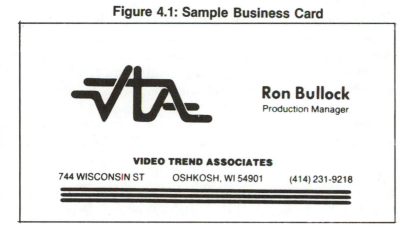

While it describes the job he does on each production, it also implies that his staff is considerably larger than it really is. He could have called himself "owner," but choosing a less egocentric job title, he has created the impression of a larger operation without crossing the line of misrepresentation.

In contrast with the essentially conservative-type business card of Video Trend Associates (VTA) is the flashy, silver-on-silver card designed for Metavision, located in Los Angeles. Industrial giants in Southern California are Metavision's target clients; therefore, the card is used to present an image of size and grandeur.

The simple business card is not so simple. It can be packed with persuasive power. Done poorly with no attention to detail, it can be just another economy print job that ends up in the circular file of your potential client. It is your first and most lasting printed impression. Give its design all the care you'd give to producing any creative project.

Your company letterhead, of course, follows the design of the business card.

**Designing a Brochure**

A brochure serves two functions. First, it is a summary of your company's capabilities. If you leave it with the client after a sales meeting, it should reinforce what you or your sales representative has said. The second function is more important. The brochure is the graphic representation of the quality and creativity that you want your company to project.

Designing and executing brochures are artistic endeavors themselves. Very few producers have the training in graphic arts to be able to do the job themselves. For this most important piece of printing, consult the specialists in your area. Look at several samples of their work. Choose the style that most closely fits your own personality and the image you want the company to have. If you can't afford to hire a professional designer, study a number of brochures, pick the one you like and try to emulate it.

The most basic brochure is a piece of 8½" x 11" paper. It can be folded in a variety of interesting ways. The most conventional flyers are folded either once lengthwise or crosswise to form a four-page pamphlet. They are printed on both sides of the page. One panel is left open with your logo and return address with space for a stamp and a recipient's address block (a self-mailer).

From this simple format, brochures explode into a complex array of everything from die-cut works of paper art, to small books. Costs for these pieces can run from less than a hundred dollars to several thousand.

The brochure for Video Trend Associates is a simple folder with a die-cut "window." Copy describing the breadth of its services is printed on individual sheets fitted into the folder. This type of brochure has the advantage of being very flexible. The individual inserts can be tailored for specific clients and needs, especially because the company does its own typesetting. The inserts are run off at an inexpensive photocopy center. With an address block on the back panel, this brochure can be sealed with a gummed tab and mailed. The total cost for a thousand of these brochures was less than $200.

One Pass, Inc. uses a similar style of brochure. While the format may be similar, the image projected is far different from that of the simple line drawings and one-color printing of VTA.

One Pass, Inc., based in San Francisco, offers a multiplicity of services described and illustrated on each of the single sheets included in its large and dramatic folder (see Figure 4.2). The brochure shows off a dazzling array of four-color photos contrasted against the stark, flat black of the printed paper stock. The package sparkles as the photos seem to float in space above the background. As a convenience, the inside of the folder holds a pre-printed Rolodex insert and one of the understated business cards for the Client Services division of the company.

**Figure 4.2: Sample Brochure**

It is a masterfully designed and printed pamphlet that implies financial success and fits the geographic locale of the company.

The brochure is perhaps the only promotional piece your client will keep. Make it of the highest quality you can afford. A tasteful piece of literature will usually find its way into a file cabinet for future reference. A tacky piece will almost always end up in the trash can, where it belongs.

## OTHER IMAGE CONSIDERATIONS

Your appearance, as well as the condition of your company vehicle, is just as important as the appearance of your printed material and your office. You become a walking, talking ad for yourself wherever you go. The car or truck you drive is also a mobile billboard, and can be an asset or a detriment.

### Projecting the Right Personal Image

Although you gain certain things by going into business for yourself, you lose a measure of freedom. If you have worked as a producer for someone else, you may have been able to shed the suit and tie at the office door—along with many responsibilities—when you went home for the day.

In some locations, especially big cities, you may still be able to retain a degree of anonymity in your personal life; in smaller areas, however, anonymity may not be possible.

As an independent businessperson, you are on deck at all times. The only places you will really be able to relax and let your hair down are in your own home and on vacation. In any small or medium-sized market, you never know when the couple at the next table in a restaurant or the guy behind you in line at the supermarket is either a possible client or someone who works for that future business account. One impression of you as irresponsible or slovenly can ruin a potential deal.

The way you dress whenever you're in public is just as important as the way you dress to call on a business account. Fashion modes vary from region to region in this country. What's in style for casual wear in Southern California might get clucks of disapproval in Michigan. Use your good common sense in this department.

Even though you are in a creative business and "creative types" are allowed to show off now and then, responsible business types are not. As a producer in business for yourself, you are caught between both worlds. Your public impression should always mirror the more conservative side.

### The Imagery of Motor Vehicles

If you can afford it, your business should have two primary vehicles—a car and a production vehicle (a truck or van). Here is some advice on choosing each.

The choice of a car requires some thought on your part. Just as realtors and salespeople, stockbrokers and interior decorators know, the company car is a status symbol. It will take you to meetings at your client's place of business. It will also take you and your clients to luncheons, as well as to shooting locations. Each time people see your car, they form impressions about you and your company.

America's complex and often irrational relationship with the automobile has been studied and written about ad nauseam. We don't just like cars. We don't just need them to get from place to place. We love them. And we choose them, typically, as external manifestations of our personalities.

With this in mind, choose the company car to reflect the personality of your business, not yourself. This choice will depend on your location and your intended market.

The car should make an understated comment on your success and stability as well. Beware of the temptation to overstate your case with an expensive luxury car. In the first place, a Mercedes or a Lincoln Continental may offend your clients' sensibilities. In the second place, the Internal Revenue Service is no longer allowing such extravagance as a legitimate business deduction. Check with your accountant to find out how far you can go in this department.

The truck or van you buy presents an opportunity to do some advertising. In addition to carrying all your equipment, the production vehicle should be a mobile billboard. The sides and rear gate present canvases for the brush of your artist or signpainter.

Production vehicles run the gamut from simple station wagons to 18-wheel monster mobile studios. Your choice will depend entirely upon the number of location shoots you anticipate and the amount and kind of equipment you will be hauling. If your firm is small and the items you will be carrying around consist of only a camera, a VTR and a light kit or two, then a large station wagon with a fold-down rear seat will probably do.

The biggest consideration when choosing a vehicle is the interior space. It must be big enough to carry all the equipment and some of the technicians to the location. It must also be air-conditioned. This is not for you, but for the equipment you will be carrying.

If you want to keep your key grip happy, purchase a vehicle with a cargo area big enough in which to stand up (see Figure 4.3). The inner walls are arranged with hooks on which power cables, light stands, etc, can be secured. Containers for breakable items are arranged for easy storage. A work bench can even be put in place for such things as on-location cable and equipment servicing and repair.

This type of configuration makes your entire operation portable. For producers with a large production volume, the pure time savings that a rig like this affords, along

**Figure 4.3: Sample Grip Truck**

with the organization and protection of your expensive equipment, can make the vehicle end up paying for itself in a very short time. A more elaborate vehicle is shown in Figure 4.4.

The type of display advertising you paint on your vehicle is critical. You can choose anything from the company logo in small gold-leaf on the door, to a full-blown master painting on the side of a large van. Talk it over with your local sign painter and include a healthy dose of your own sense of good taste.

A promotional vehicle can and will attract attention. It can make your company look larger than life and enhance your image of strength and durability in the community. If you aren't careful with your design, however, the vehicle can be a detriment. Finally, since the vehicle is your company's image in motion, choose a careful, good driver.

## Community Service

Many independent producers donate their time to worthy causes. Others feel it is essential to become a part of the community. You meet many businesspeople in service groups, and people who know each other and socialize together also tend to do business with each other. Very simply, a member of one of your service groups who wishes to invest in a video production will probably be more inclined to do business with your company than with a producer he doesn't know.

If you wish to give something back to the community in which you live, you can offer your service as a video producer.

**Figure 4.4: Production Vehicle for Large-Budget Location Projects**

Photo courtesy of Image Devices International, Atlanta and Miami.

There are literally hundreds of worthy causes in this country. Here are just a few ways in which you can offer your expertise as a community service.

Choose a local organization for which you can make TV public service announcements (PSAs). Large national charities such as the United Way and the Red Cross have major PSA campaigns. However, your local Boy's Club, Youth Hockey Association, Church Service League or Big Brothers and Sisters organization do not. Area television stations will, in most cases, prefer to run locally originated PSAs because doing so helps them fulfill the community service function required by the Federal Communications Commission. The only investment for most PSAs is a day or so of your time. The payback in terms of the good will you generate is incalculable.

You can also pick out one of the youth groups in your area and offer a seminar in video production at your facility. Let the kids develop a script and shoot a project under your supervision. You may be astounded by the creativity and visual sophistication that most young people have today. The end product can run, again, on one of your area stations as a public service presented by them and you.

Try contacting your community social services administration. Various states have different names for this organization, which handles such things as residential care for the developmentally disabled, programs for the physically and mentally handicapped, etc. Frequently, this type of organization has a need for such things as training aids for new personnel and progress reports for state and country funding agencies, both of which can be accomplished with video. You can provide a much-needed volunteer service and have the double benefit of helping those who can't help themselves while reducing the burden on the taxpayer for production services.

With very little research and effort, you can find many more community service projects. Volunteering in these areas will certainly enhance your standing and prestige in the community. Always project the positive image of a shaper and a doer. According to the producers from whom these cases are drawn, you can be assured of an additional benefit—you will feel very, very good about yourself.

## MAKE USE OF THE MEDIA

Laypeople feel that video production is a profession with a built-in "glitter-factor." You know that producing video is not glamorous; it is 95% sweat and drudgery and 5% pure joy. That 5% is what keeps you going.

Most "outsiders" see producers and productions as emanating from a world of fantasy and fun. Knowing this fact means that, for once, you can turn the media to your own ends.

### On-Location Shooting

Most of us in the field have had the experience of being out on a city street with a crew. We know that a crew always draws a crowd of spectators. You may be doing the most pedestrian kind of shot, an exterior of a model entering a shoe store, for example. But the sight of the camera, the crew and a couple of Colortran "kickers" puts you in the same league as MGM or 20th Century-Fox, as far as the person-in-the-street is concerned. This is especially true if you are in a small or medium-sized town. Since productions in these areas are not as common as they are in cities like Los Angeles or New York, your mere presence on those city streets can be like a media event.

Many television stations use a great deal of feature material, and not all hard news, on their evening newscasts. This is especially true in small and medium-sized markets, where there may not be that much hard news in a day. Therefore, local stations may welcome your on-location video production as a feature story on the evening news. If this opportunity arises, take advantage of the exposure and cooperate with the station.

### The Press Release

To let the media know what you are doing, use the press release—a basic tool of public relations. Write it from the point of view of the third person and use language that reads well aloud (i.e., write it the way people speak). If your press release reads well, your local announcer or editor does not have to bother with a rewrite and you have a better chance of having the piece aired. Don't be afraid to make it sound a little more glittery than the project may seem to you. Remember that to the other folks in your town, it's really show biz!

The first paragraph of the press release is called the *lead* and usually includes the "five Ws and an H" of journalism. These are: who, what, where, when, why and how. All significant information is included in the lead because that paragraph is typically the only one that is read or printed in its entirety.

In subsequent paragraphs, expand on the information. Include colorful details and quotes from those who are involved in the story. When you want to say something directly, quote yourself as if you had been interviewed by the writer. Conclude the press release with a recap of the most important information.

Look at the example of a press release used by the author to receive media coverage for his company and to recruit some local talent for a major commercial production (Figure 4.5). Note that it states, "For Immediate Dissemination." Be sure to include this, or a similar phrase, unless you wish it held for a specific date.

This sample press release follows basic journalistic rules. All of the essential information is contained in the first paragraph, the lead. In the second paragraph we expand. We name the sponsor and, in the interview form that follows, the sponsor's executive director tells us the reason for the project. We put the sponsor above the name of the production company for a couple of reasons.

First, it is good form to let the sponsor have top billing. Second, it is always good publicity to be connected with an altruistic foundation and project.

We blow our own horn lower in the story because we don't want to look as if we are seeking publicity. There is a perilously fine line between newsworthiness and advertising. If the news director or editor feels you are trying to advertise, your press release may not be aired.

Do not overdo this public relations tool. If you get in the habit of sending out press releases each time you do a production, your welcome at the local newspaper and local television and radio stations will be short-lived. Three or four times a year, though, when you have something unique, fire up the typewriter. You may very well find friends within the media who appreciate your occasional press releases because they fill some space for them.

## DIRECT MAIL

Microcomputers have opened the way for even the smallest business to enter the direct mail method of promotion. You can buy a microcomputer with a word processing program suitable for generating form letters and mailing lists for less then $1000. As competition in the field continues, the price of a good machine will fall even more.

Computers are also becoming more and more user-friendly so that if you can type, you can use a word processor. Word processing makes scriptwriting—with the frequent changes required by clients—a real pleasure. With programs such as spreadsheets and databases, your computer can help you with budgets, cost analyses, accounting and a host of other details.

**Figure 4.5: Sample Press Release**

PRESS RELEASE                              For Immediate Dissemination

Date:_____

HEADLINE:

**LOCAL PRODUCTION FIRM LANDS NATIONAL CONTRACT**

An Oshkosh production company has been chosen to make a series of national television spots. The theme of the public service announcements is "Aging in America". The California-based sponsor of the series chose the Oshkosh firm after meeting here last week to review the facilities and the budget. The contract was awarded on the basis of a low bid and a high quality demonstration reel of past projects. The first two spots will feature Wisconsin aviation pioneer Steve Wittman and famous author, Ray Bradbury. Production of the series will begin here with Wittman in March.

The sponsor is The Elvirita Lewis Foundation headquartered in Soquel California. The purpose of the Foundation is threefold, according to its Executive Director, Steven W. Brummel.

"We provide funding to programs for the elderly", he said. "We believe that older people are resources, not drains on society. We fund several projects a year where older people run the programs for themselves."

"The second part of our operation is to lobby on behalf of elder citizens in the areas of health and nutrition", he went on. "We are in touch with other foundations and government institutions worldwide on these subjects".

"Finally", he concluded, "we are trying actively to change America's attitudes about aging and the aged. That's the reason

**Figure 4.5: Sample Press Release (Cont.)**

for the first in our series of television spots. We want to show older people being active and productive long past the point where many younger people feel they can be".

Dr. Bob Jacobs, Executive Producer of the local production house called The Media Ranch, expressed his enthusiasm for the project.

"We took this on at essentially no profit", he said, "because we believe very strongly in the message. When I told Mr. Brummel about Steve Wittman's willingness to do a spot for us, I think it helped clinch the deal!".

Steve Wittman, for whom the Winnebago County Airport is named, still builds, flies and races airplanes. Wittman is now in his 80's.

"That's exactly the message we want to get across", Brummel verified. "People in normal good health can do just about anything they've always done no matter how old they are".

The production will start here in March and move to Los Angeles for the Ray Bradbury segment in June. Release of the spots on a national scale is slated for December. There will be minor roles for extras and background characters in Oshkosh. Interested performers should call The Media Ranch for tryouts.

The major advantage of computer-assisted direct mail promotion is that you can send a personal letter to a large number of people. Each letter will look as though it has been typed individually, provided you have a letter-quality printer with your word processor. The dot-matrix printing system is much faster, but is a dead giveaway that you are sending a mass mailing.

Figure 4.6 is an example of a promotion letter sent out by Video Trend Associates. It is an introduction to the general services VTA offers and a pitch for its new Video Real Estate operation (discussed in detail in Chapter 5).

The text of the letter is filed on a floppy disk. The address blocks to more than 100 real estate firms in the state are filed on a separate disk which then feeds the information to the printer. The name of the person or firm from the address block is programmed to print in the appropriate spot in the form letter. This simple device makes it look as though the letter was written specifically for this individual.

The cost per letter is computed simply by adding the cost per sheet of letterhead, the printer ribbon used and the time it took you to write the original and punch it into the computer.

Once you have established a mailing list of potential clients and filed them on a floppy disk, make personalized mailings a regular part of your promotional efforts. Here are a few direct mail ideas in use with other producers.

- Send a monthly newsletter. Tell your readers about new equipment, problems that you have solved for other clients, etc. Make this chatty and add some humor if you can. You are not making a sales pitch; you are simply making clients feel as if you value them as part of the "family."
- Send personal holiday greetings. Include some quips about you, the staff and the company in general.
- Pass on helpful hints that you have picked up from trade journals.
- Write about some new technique that you have picked up. Tell the clients why it can help in the next project you do for them.

With a little imagination you can invent dozens of ways to use direct mail as a first-line public relations and sales tool.

## UNCONVENTIONAL TACTICS

Promotion can be a silly business. The limits are really only defined by your sense of propriety and an awareness of what your clients will tolerate.

Gimmicks can range from giveaways such as pens and matchbook covers to some truly outrageous stunts. One producer, for example, has gone to the extreme of staging phony shoots in proximity to clients he hopes to land. He sets up lights, hires actors and pretends to be making a major production. A few days later he makes his sales call and, in most cases, the client recalls having seen him "at work."

**Figure 4.6: Promotion Letter**

VIDEO TREND ASSOCIATES

Century 21/Paul Schmidt Realty
325 Pearl Avenue
Oshkosh, WI 54901                          December 3, 1985

Dear Paul Schmidt:

     We're writing to introduce ourselves and to tell you about a
new service of ours.

     Video Trend Associates began in a very small way six years
ago, believing that there existed a need for a high-quality
alternative to the production services of the three Green Bay
television stations. While much of their production is of a high
standard, their overhead costs demand an equally high price.

     We have been able to keep overhead costs low while being
more than competitive in terms of production values. Some of our
satisfied clients include The Federal Aviation Administration,
The  Experimental Aircraft Association, Air Wisconsin and The
Marcus Corporation in  Milwaukee.

     We offer a complete range of services in 3/4-inch
professional video production and post-production. We can also
reproduce works in 1/2-inch Beta and VHS formats.

     Of special interest to Century 21/Paul Schmidt Realty is our
new "Video Real Estate" Service. Professional tours of your
listed properties, showing their best features, can be recorded
on video and shown to your potential clients in the comfort and
privacy of your office. This saves you both travel time and
money. It is especially effective in winter months which are
traditionally bad for your business.

     This service is already widely used on both the East and
West Coasts with great success. Video Trend Associates would like
to join Paul Schmidt in bringing video sales of real estate to
the Fox Valley of Wisconsin. We would be happy to show you how
remarkably inexpensive and very effective a video production can
be.

     For an appointment and free screening, please call us today.

Cordially,

*Ronald J. Bullock*
Ronald J. Bullock
Production Manager

RJB/ac

744 WISCONSIN ST. ● OSHKOSH, WI 54901 ● (414) 231-9218

Most producers avoid gimmickry. It comes close to going over the ethical demarcation line of sound business practice. On the other hand, production is a highly competitive and creative line of work. And a well-controlled and harmless stunt may occasionally mean the difference between you and someone else landing a contract.

## SUMMARY

As in any other business, video producers are in competition for a limited amount of customers and a limited amount of dollars. We have discussed a number of methods of promotion and advertising, many of which are currently in use by other producers. The type of promotion technique you use is unimportant, but the decision to promote at all is crucial.

If you feel that you will not excel in the areas of self-promotion, you can use the services of an industrial advertising or public relations agency which specializes in promoting businesses rather than products. It will probably cost less to do your own promotion; however, advertising is important enough to the success of your business to pay professionals if you feel you do not possess the necessary expertise.

## NOTES

1. For help in designing your own advertisements, see Albert C. Book and Dennis Schick, *Fundamentals of Copy and Layout* (Chicago: Crain Books, 1984).
2. Ron Bullock, interview with author, February, 1985.

# 5 Finding and Developing Markets

The typical independent producer develops one or two primary accounts that serve as the foundation for his or her business. These are called bread-and-butter accounts and they typically occupy 60% to 70% of the producer's time.

Once you have secured these types of accounts and your major bills are being paid on time, you have two options. You can sit back, relax and feel as if you've got it made, or you can pursue a series of secondary accounts. If you're smart and want to stay in business, do the latter. In this chapter, we will discuss both the bread-and-butter accounts and the secondary accounts—how to get them, how to keep them and what they can mean to your bank account.

## THE CORPORATE CLIENT

Major corporations spend billions of dollars collectively every year on audiovisual productions. The projects include multimedia presentations for annual sales meetings, training, intracorporate communications, sales promotion and new product research and development, as well as television spot commercials and special programs for broadcast.

Many large companies have in-house production units that handle their day-to-day needs. Some of these, however, employ outside free-lance help from time to time on a per-project basis. A larger number of corporations have no audiovisual production capability in-house, and instead contract exclusively with outside agencies for their production work. Write letters of inquiry to those companies with which you would like to work. Ask if—and to what extent—they subcontract for production work and how you can get on the bid list.

The world of corporate audiovisual production offers a multitude of possibilities. For each giant that relies solely on its own production capability, there are dozens of smaller firms that hold interesting and potentially profitable opportunities for the independent producer. Landing one of these as your primary account can put your business on a sound, relatively stable base.

## The Game Plan

In order to land a major corporate account, you must develop a game plan and follow it. Although no one can tell you exactly how to go about doing this, we have provided certain techniques and tips throughout this book that can be used to land almost any type of account. In addition, three case studies at the end of this chapter—as well as the words of wisdom from Hollywood producers Saul Bass, Henry Winkler and Saul Turtletaub that appear in Appendix D—should prove helpful.

When meeting with potential corporate clients, remember that you are dealing with people who probably have experience in the field. Treat them accordingly. While you are selling your idea, be certain that you are also being an effective listener, concentrating not only on what is being said, but on what is being implied. By listening carefully, you can often pick up on problems the prospect has had in the past or learn of one or two issues that might lend themselves to a video solution. Keep notes of these and ask direct questions relating to what you've been told.

Remember that corporations are not impersonal, calculating entities; they are made up of human beings. Approach them with that in mind at all times. Often, they have needs for services like yours. Approach them in a businesslike manner with a solid proposition at a fair price. Deliver it on time and on budget and make your word your bond. If you run into a brick wall, move on to another prospect. If you follow these few guidelines, however, you have a good chance of enjoying a long and profitable relationship with the corporation of your choice.

## The Proposal

You will have to submit a written proposal to a corporation for your production. This will follow one or two meetings with the client, who has agreed that there is a problem in the corporation and that a video solution is needed. Be aware that there are standards of practice for you to follow here. Remember that your time is valuable; do not give away too much of it for free. The informational meetings that you conduct and the research that you do into the nature of the problem are your investments of time. The written proposal is the final investment you should be expected to make without remuneration.

In the proposal, you recap the problem. Simply stated, you tell them what they told you. They will agree with you. Then tell them that a video solution will solve the problem. They will agree. Then tell the clients that you have a script idea in mind and you will be happy to work on it with them. Then give them a rough estimate of the cost, telling them that you cannot give them a detailed production budget until the script has been written. Then tell them how much you will charge to write the script.

Many producers have found themselves so eager to land an account that they do the script on spec, believing that this will impress the clients. Keep in mind that you are a professional and that your time and talent have value. Corporations understand this and play the game accordingly. If your proposal is not strong enough to sell the prospect and to land a contract that will pay for the rest of the stages, move on to another prospect. (See the case studies at the end of this chapter for more information.)

In Appendix E, the proposal that the author used to land an account with Mercury Marine is reproduced. You can see each of the steps we've outlined in the proposal. The goal is a single recruiting production for the product engineering department. The preliminary research with the prospect is detailed. The problem is discussed and the solution proposed. The estimated budget is outlined along with the terms of the production contract. Feel free to use this example as a format of your own. (But don't pitch it to Mercury Marine!)

## THE ADVERTISING AGENCY

In recent years a rash of advertising agencies has been springing up in small and medium-sized towns and cities across the country. Licensing practices for these agencies vary so widely from state to state that there is little agreement on what constitutes an advertising agency.

Often an agency is a one-person shop with three or four clients for whom the owner does everything—from writing and laying out newspaper ads to appearing in local television spots. Exercise great caution in dealing with operations like this, because some have reputations for not paying their bills. This is not to imply that you should avoid all small agencies, but you must use care and good judgment when evaluating them.

To be on the safe side, do business with agencies that are members of the American Association of Advertising Agencies, Inc. (AAAA or Four A).[1] When you deal with a Four A agency, you can be sure that it has been around long enough to prove that it can study the advertiser's business and products, analyze the market, form sound judgments, give constructive advice and render an adequate quantity and quality of service.

There are some legitimate agencies that fall between the local scam artist and BBDO. Some perfectly reputable agencies may have been in business too short a time to qualify for Four A membership. Some may have no need to join the association because they work with a select client list. Before striking a business deal with an agency, consider the following.

A respectable agency will be happy to provide you with a client list. It is more likely to operate out of an office than out of someone's home. It will probably have more than one employee—at the very least, a marketing research person, a graphics specialist and a media production person should be on board.

Before approaching an agency you know little about, check with your local television stations, as many small agencies use these types of production facilities. Ask some straightforward questions regarding the agency's credit history. Check also with your Better Business Bureau to find out if it has any complaints on file and ask the local Chamber of Commerce about the agency. Most respectable businesses (like yours, for example) are members of the Chamber and, therefore, they have agreed to abide by certain sets of ethics and standards of practice.

Beyond this, let your instincts be your guide. If the agency representative seems slippery or avoids answering several questions, back away.

Once you're satisfied that the agency is solid, follow essentially the procedure that was outlined in Chapter 1 under ''Selling Your Idea.'' Some elements of selling an idea and landing a contract are the same regardless of the potential client; other techniques must be tailored to a specific prospect. Following are some characteristics of a typical advertising agency—they are very different from those of a typical corporation. It is essential to keep each industry's individual characteristics in mind when trying to land a contract.

First, the agency is on your side of the fence. Its employees know all about production and already have the client. Consider the agency as an intermediary. You don't have to sell an agency on the benefits of video; therefore, your job is much easier.

You will probably be in contact with either the creative director or one of the in-house media producers at the agency. These people often have years of experience in or around production. They are usually creative people, much like yourself. Because of this similarity, it should be a bit easier to establish a rapport with them than it is to get on the same wavelength with corporate presidents. It also means, however, that you can't put anything over on them, so don't try!

## Costs

Be prepared to demonstrate that you can do high-quality work. Since most agencies are already doing business with other producers, you probably will have to offer a lower production cost and be able to prove this on paper in order to take a share of this business for yourself. Point out your lower overhead, for instance. Since agencies almost always operate on a percentage of the clients' media buys, they are interested in getting the most out of their production dollars without sacrificing quality. In some cases the client is assessed a production charge. It makes the agency look better if the figure comes in lower than planned.

Production budgets for television spot commercials vary according to the size of the agency, the size of the client and the nature of the spot. In small and medium-sized markets, a typical production cost for a 30-second video commercial for a small business runs from $500 to $1000. Fees for film-to-tape commercials run from $2000 to $5000, depending on whether the film stock is 16mm or 35mm.

From these minimum fees, the costs can go up to the astronomical. (One spot by a major electronics firm for Super Bowl XX had a production budget of $1,000,000!) Factors that determine costs include talent (AFTRA and/or SAG talent is very expensive), special visual effects, length of the spot and whether you are doing a series or a one-shot. The agency will generally indicate these factors to you when they call for bids.

## Other Factors

In addition to cost, a typical medium-sized agency will look at the following when deciding which producers to place on its bid list:

- Creativity: It wants to know that you're capable of pulling off a range of productions, from the humorous to the heart-tug. (The demo reel is covered in detail in Chapter 1.)
- Technical competence: It pays close attention to lighting, audio, editing, acting and any special-effects skills you have.
- Past credits: Nothing impresses more than past success.
- Time allotment: Will you be able to put in full-time and overtime hours, and whatever else is necessary to deliver the project on time?
- Staffing: Be prepared to introduce the agency to your videographer, editor, artist, etc. It is difficult for one person alone to hook into an agency account.

## ALTERNATE MARKETS

The corporate account and the advertising agency are traditional markets for the independent producer. For the energetic entrepreneur, there are also secondary markets that offer profit potential. Here are some of them for you to consider.

### News Stringing

Every broadcast news operation—from the networks to your local TV station—is a potential client for the producer who keeps an ear tuned to an emergency band radio scanner. The scanner picks up the police, sheriff, fire department and other emergency or official frequencies of the radio band. If you hear what sounds like a major breaking news story, you or your videographer can grab the camera and VTR and go to the scene of the story. While all news operations use the scanner, stories are often over before news teams can get there to cover them. If you have the only coverage, you can sell it to one or a number of stations.

Almost anyone with a broadcast-quality camera and VTR can become an official "stringer." This means that a news service or station has given you a press card, which affords you the same rights and privileges as a full-time photographer covering the news. Get in touch with your local stations and the networks if you want to become a stringer. Bear in mind that members of the press are often denied access to hazardous situations. The card, however, does distinguish you from the curious onlooker.

In hot situations you will have to fight for your right to shoot. The members of the press are a scrappy bunch of people; don't be surprised if you find yourself in shouting and shoving matches when trying to get your camera to the front line of a major event. And don't take it personally. It's just the nature of the game. You can be polite, wait your turn and possibly miss the action. Or you can assert yourself and get the shot. There is no in-between.

News stringing requires you to keep the batteries charged on your camera and VTR at all times, and good stringers always have their gear packed in the trunk of the car. These people often get big breaks because they are prepared.

Fred Schuh, an international free-lancer who now travels the world at network expense shooting the news, got his big break this way. He was driving to dinner over a causeway in Tampa Bay when a huge freighter rammed the bridge, knocking out an entire section of it. Fred was not only the first one on the scene, he was the only one on the scene with a camera and VTR. Schuh's coverage of this major accident was sold to all three networks and helped establish him firmly as a top free-lancer.

## Music

There are at least 100,000 musical bands in this country. They range from preadolescents with a three-chord guitar ability and a five-tune repertoire to top-forty recording artists with major label contracts. Thanks to the phenomenal success of cable television's MTV (Music Television) as an alternative to radio, all of these bands are potential music video clients.

Providing a music video is really quite simple. The band records a song. The song is lip-synced in playback and you shoot visuals of the band members (and any of a stunning variety of other characters and images) to go with the song. Then you edit all of this visual material into a series of flashy pieces cut to the beat of the music. A typical two- or three-minute pop video will have over 100 separate cuts in it.

Finding clients is a fairly simple matter. In between the preadolescents and Michael Jackson are the serious professional or semiprofessional bands. These young musicians play in nightclubs and on college campuses, and finance their own recordings hoping to attract the attention of a major label. They have money to invest in their careers. A nearby audio recording studio is a good place to find these musicians and advertise your services.[2]

Check, too, with local colleges and universities. Most have booking services for music and other entertainment. They can put you in touch with the national publications distributed to the groups that play the college circuit.

Finally, look in your newspaper's entertainment section. Find out where the live music is and go see the groups in person. Present a flyer advertising your music video service. Word of mouth will spread among the musicians if you give a good product at a fair price.

The Basement Tape Competition, sponsored by MTV, is one impetus for the swelling number of groups that want to have music videos produced. MTV solicits the submissions from amateur, semiprofessional and professional bands; the only restriction is that the band may never have had a recording contract. Each month winners are chosen to go to the finals. The big winner receives a recording contract. There are no limitations on the videos themselves. And if you produce a winning one, you can count on being deluged with more business than you might want to handle.

A minimum figure charged by most producers for a music video is $1000. From there, the price goes up according to the resources of the musicians and the level of technical difficulty they desire.

## Weddings

Luckily for video producers, the home videocassette revolution is well under way. Laypersons now speak of Beta and VHS, and of programmable features, etc., as if they were old hands in the video business. This wonderful turn of events opens up more markets for you—the wedding is one to consider.

Video producers are rapidly complementing—if not supplanting—the traditional still photographer at weddings. For some producers, the wedding season from June through August has become a primary source of income.

Advertise your services at the local bridal shop. A brochure describing just what you will do is a good handout. You can also place ads in bridal magazines if you want to attract business from a larger region. This is one area where you can benefit from some television spot advertising, too. Local stations frequently have very low spot rates on late-night programming in nonmetropolitan areas. Your sales pitch should be that you will deliver a "living, moving reminder of this most important of days."

Most wedding videos are very simple to do. Choose a piece of music long enough to cover 5-to-10 minutes. Lay it down on the audio track of a blank tape. Then, simply insert video edits in time to the music, using fades and dissolves to enhance the romantic feeling. Follow the wedding party from arrival at the house of worship, through the ceremony and into the reception. Remember the standard wedding photos: the bride and groom in front of the altar, kissing, cutting the cake and so forth.

Be aware that most music is copyrighted. Many favorite wedding songs require a license to reproduce. If you use a copyrighted song, obtain a license from the publisher to do so. The licensing fees for noncommercial, nonbroadcast use are usually very modest. Contact the American Society of Composers, Authors and Publishers (ASCAP) or Broadcast Music, Inc. (BMI).

## Real Estate

The real estate video market is growing at a rapid pace, especially in the Midwest and northern parts of the United States. When weather is inclement, or valuable properties are buried in snow, you can help local realtors make sales in the warm comfort of their office.

The process is very simple. You produce video tours of homes that the realtor has listed. The realtor can do commentary as you go, pointing out the features and highlights of the property. These tapes are so simple that often little or no editing is required. It's very similar shooting news; include a ''B'' roll of fine details that you might wish to edit in later.

The realtor advertises this video tour of homes as a nifty service for both buyers and sellers, as the tapes give homes added exposure with less intrusion on the life of the seller, and they save potential buyers the trouble of traveling to see houses that they would never be interested in.

You update the reel biweekly or monthly as properties are sold and new ones are added. Fees for video real estate services can be negotiated on a flat fee or on a monthly retainer basis. An average of $500 to $1000 a month is customary for the latter. When you point out the savings to the realtor in terms of expensive color photographs, brochures, wasted house visits, etc., you have a fine selling tool. For those offices that do not have a VCR, many producers offer a package that includes the VCR and maintenance.

## Other Possible Markets

The number of potential secondary markets for the creative independent video producer is almost infinite. Keep your mind open to new ideas. A few innovative suggestions follow.

Insurance companies recommend that holders of homeowner's policies take careful inventory of all valuable possessions. They also advise that photographs of these belongings be kept for identification purposes. The entrepreneurial independent producer should be able to see how these facts can mean increased business.

Explore the notion of video yearbooks with your local high schools. Producing a yearbook on tape can be time-consuming, as you will have to be involved throughout the school year covering major events, homecoming, student leaders and the graduating class. The music score is the high school alma mater, fight song and so forth. On one day, just as the printed yearbook does photos, you do short spots of each senior making a comment on life at good old Washington High. The video yearbook has caught on in many affluent areas of the country.

Many offices—including those of dentists, doctors and others—are turning to mood tapes to keep patients, clients and employees relaxed. The types of visual images included in these tapes depend on the mood that your client wishes to create.

Make a sample reel of the sorts of things you'd like to watch and hear at the dentist's office and try to sell it. Production costs for mood tapes are usually quite low.

We could go on here listing possibilities for the creative producer. Use this list as a starting point. Don't get locked in on two or three potential markets and ignore all else; the world is an electronic playground—use it to your advantage.

## CASE STUDIES

The following case studies reveal how three independent producers tried to land and keep major corporate accounts. Two were successful, one failed entirely and went bankrupt. The case studies are all true, although names and places have been changed.

### The One-Account Jump

Spec Video, Inc. was a medium-sized midwestern production house. It began in the late 1950s as the graphics arm of a small ad agency. Its director at the time, Bill Miller, did a small amount of 16mm film work for a few clients, but specialized in technical drawings and illustrations for the firm's mostly print-oriented accounts. One of these was the Garden Corp.

Garden was the manufacturer of a line of power garden tools, riding lawn mowers and snow throwers. Garden was comfortably positioned as the leader in its field; its prices were high but so was the quality of its products. The company was controlled by the founder's family, all of whom were very conservative. Spec Video's main job was to produce the annual catalog and equipment brochures; an occasional film was produced to demonstrate the new line to dealers each year.

The ad agency lost the account after a number of years. At the same time, Miller decided to buy out the production wing of the agency. He took on Garden as his main account and Spec Video, Inc. was born as an independent production house.

Soon after, the competition heated up for Garden as other manufacturers got into the rapidly growing power garden tool, riding lawn/garden tractor and snow thrower market. Large firms like John Deere, Toro and even Honda launched major television campaigns to steal the position once dominated by Garden.

The company went with a larger ad agency in Ohio, but retained Spec Video as its production house. Though it had started primarily as a graphics producer, Miller's company found itself now doing large-scale television spot commercial production for the full line of Garden products. The staff swelled to eight full-time employees as demand grew for a variety of services ranging from technical illustrations and brochures to multimedia presentations. And it all hinged on one major account.

Up to this point, this producer had been very fortunate. He had landed a lucrative corporate account and understood that to keep it, he had to try to give it a full range of services. He could not just stick with producing film and video. He established a personal relationship with the people at Garden and enjoyed their corporate loyalty to him for a long time. Corporate loyalty, however, can be fickle.

When Garden's management saw a spot commercial featuring stop frame animation of a product, it decided that the company needed a change in its prosaic image. Garden suddenly dropped Spec Video as the producer of its spot commercials and went with a Chicago outfit whose specialty was animation. The cost of the new spots was enormously more than the fees that Spec Video had been charging. The cost to the producer for losing this single area of the Garden account was nearly fatal.

While it retained much of the graphics and multimedia work for Garden, Spec Video reeled under the impact of having placed all its eggs in one basket, and then losing some of them. After enjoying nearly two decades of sustained growth and prosperity primarily from this one corporate account, the producer had to shift gears and make a concerted effort to go after other accounts to pump up the deflated bank book. Soon after this turning point, Miller sold the company.

The lesson to be learned here should not be forgotten. Never rest on one account, no matter how solid the foundation seems to be.

## A Case in Point

Jack and Bill were young producers who had just left the United States Air Force, where they had worked for five years as officers in television and film production. They had saved their money and, with a small boost from the Veteran's Administration, opened a production company north of Los Angeles and close to major industry in Southern California.

They were complementary partners; Jack was strong in production, Bill in graphics and sales. Each had established good contacts with industry while they were still in the service. Their town had some modest industrial firms whose needs they could fill at a lower cost than could any producer in Los Angeles.

They went after a major account, a nationwide chain of fast-food restaurants we will call Joe's Café. Jack and Bill knew that the fast-food business had a high turnover rate in waitresses. They had done some research and found that Joe's Café was spending a relatively high percentage of its income on training. With this knowledge, they devised a proposal to provide video training to Joe's waitresses. Their package would include programming and video playback units supplied on a rotational basis to the restaurants around the country.

They spent a month in preparation (that included making up charts and graphs and designing a tasteful brochure), and then Bill made an appointment with the president of the restaurant chain.

The presentation went well. Joe, the president of the chain, told Bill that it was very timely since he and his staff had just been considering the subject of training. He wondered if Bill and Jack would be interested in putting together an entire package that would also include a handbook and other materials for waitresses. Bill was ecstatic.

At Joe's request, Bill left the entire presentation package, including storyboards for the training video that he had completed in an effort to show just how ready he and Jack were to roll on this thing. Then he went back to the office to prepare, at Joe's request, a sample of the training manual.

Several meetings between Joe, Bill and Jack ensued. At each one, Joe expressed his delight with the work that the two producers were submitting. By the eighth meeting in as many weeks, Bill and Jack had presented the complete outline for the manual, the script for the training video, several pieces of finished artwork, a shooting schedule, the final storyboard and even some sketches for new menu art that Bill had done in his enthusiasm. All work on trying to land other accounts had stopped.

Joe was absolutely thrilled with Bill and Jack. He told them so many times. He also told them that the contract between them was "being worked on" and that they could expect the go-ahead and a check for a substantial down payment on the production budget at the next meeting.

The next meeting never came. First, Joe was "called out of town." When he came back he was "unavailable" to answer telephone calls from Jack and Bill. Rumors began to spread that Joe's Café was in trouble of some sort. Needless to say, Jack and Bill's production company was in terrible trouble by this time.

The two men, believing with absolute certainty that Joe's was going to be their bread and butter, had eaten through their marginal capitalization almost entirely. The one or two other clients for whom they had done some small jobs had turned to other producers. Unable to get back in touch with Joe or anyone else in the organization, Jack and Bill folded their dreams along with their tent, paid their creditors with the last of their money, closed up shop and went back to work for another company in Los Angeles.

The final salt in their wounds came when they found that their script and storyboard had indeed been used by Joe's Café, which had never been in trouble at all. The two producers had literally been robbed.

Jack and Bill made some very big errors. First, they assumed that a corporation would be honest. They were, in short, suckered. Second, they presented work, time after time, without asking for payment and without having a written contract. Since the meetings were always with Joe, they had no way of proving after the fact that the training video that another company produced for Joe's Café was actually theirs. They had not copyrighted it and, in court, it would have been their word against Joe's.

Finally, they had lost sight of their primary goal, which was to produce a training tape. Writing manuals and doing cover art for menus obscured their vision of what they had set out to do in the first place. And this loss of vision proved fatal to their company.

## The Front Desk

Don worked for several years as a musician and a free-lance photographer. He had a good financial cushion and a growing interest in video. So one day he replaced the movie camera with a video package, including a modest editing setup. He became hooked, and thus the next logical step for him was to go into business. We will call his company Bootstrap Productions.

His local university had a good program in Radio-TV-Film. The campus also had a fine college of business. Don enrolled in a series of courses that he tailored to fit his needs. Concurrently he managed to develop a growing list of small clients. His on-the-job training approach soon had his equipment paid off and he added more sophisticated gear. By the time he was finished with his courses at the university, his company was almost paying for itself.

Don recognized that to get his company on a sound footing, he had to get a large corporate account. Because he disliked selling, he also wanted to find a salesperson. He learned of a young woman named Janet who was in the graduate program at the college. She had enough technical knowledge of the field to speak with authority and she enjoyed sales and marketing. Janet became his associate producer.

Janet had been working part time at a hotel that was part of a large chain. She knew that they needed help in training the staff and suggested approaching the corporate headquarters, even though they already had an outside production company. Because of the low overhead of Bootstrap Productions, Janet felt that they could undercut the competition. She and Don prepared a modest proposal for a training video, which included a demo reel of Don's work.

The president of the corporation was very impressed with the presentation and agreed to let Don and Janet take a crack at a short training tape for the hotel's housekeeping staff. The go-ahead was based in large part on the low cost quoted by Bootstrap. The budget had been whittled down to no more than a break-even deal if Don and Janet could stick to the three-day shooting schedule. Even one extra day would put them in the red.

They took the risk of a relatively small financial loss to get their feet in the door. And taking this chance paid off. The shoot went smoothly, the tape was finished and the client was happy as a clam about it.

A couple of months later, Don and Janet were invited to bid on a new contract for the corporation. The president called them in to discuss his need for a similar training tape for his front desk personnel. This time, he also wanted a written training manual thrown in.

Having literally given away the first project, Bootstrap Productions was now in a serious situation. Neither Don nor Janet was competent at technical writing—certainly not competent enough to do a complex training manual from scratch. The corporation was putting on a squeeze play as well. While the president knew that the first production had been a nonprofit trial, he made it clear that he expected this bid to come in substantially lower than any others he had been offered. Don and Janet knew that the price they quoted for the video production would be accepted, if they could figure out a way to come through with the manual.

Don and Janet agreed to try to find someone who could do the manual. The president, knowing that he was, in fact, getting a good deal on the video, agreed to let Bootstrap Productions have a few days to put together a separate proposal for the manual. Don knew a woman who was a good technical writer, and subcontracted with her to do the writing. He then added a 50% markup on the writing as a profit to his company. The figure he presented to the corporation was lower by several percentage points than the next bid, even including the Bootstrap markup.

The manual was completed on time, the training tape based on the manual was finished on budget and Bootstrap Productions landed its first ongoing major corporate account as a result.

It is clear that no matter how badly you want or need a corporate account, you have to stick to your game plan. In this case, the manual did not fit into the plan. It would have been wrong, as Jack and Bill discovered, to agree to it.

Don and Janet understood two things: their limitations and their financial needs. While both of them were tempted to capitulate and give the corporation something for nothing, they understood the first rule of good business. You can't stay in it if you don't make a profit. By standing their ground on the amount of their bid, they won the respect of the corporate president. By delivering what they agreed to deliver for the price quoted, they won his continued business.

## CONCLUSION

In this chapter we have discussed some of the nuts and bolts of getting your accounts receivable column moving toward the black. All the case studies have come from actual experiences of other independent producers. Your circumstances may or may not be similar to theirs. From their stories, however, you should be able to glean information that will help you avoid the snares and traps that they have gone through and stimulate your mind into productive channels of current and future profits.

## NOTES

1. For an annual listing of Four A members, write to AAAA, 666 Third Ave., New York, NY 10017 for *Roster and Organization*. This directory is available free of charge.

2. For a list of many U.S. audio production houses, see *The National Register*. It is available from the National Register Publishing Co., 5201 Old Orchard Rd., Skokie, IL 60077.

# 6 Procedures and Practicalities

In previous chapters we covered the business of business in detail. In this section we will cover several standardized practices that relate specifically to independent video production.

## ESTABLISHING A RATE CARD

In the retail business, pricing a shovel or a pair of slippers is easy. Merchants figure the cost of the product, factor in their overhead, compute a figure for losses due to theft and damage and add on a reasonable profit margin.

In some ways, the independent producer does the same thing. You know, with some certainty, what equipment, tape, talent, crew and post-production costs are going to be at the outset of a project. You may be lacking guidelines, however, for the cost of your most valuable assests: your own training, skill and time. These are the same intangibles for which people willingly pay doctors, lawyers and psychiatrists. Yet many of those same people fail to recognize what costs so much in a video production.

Since we are not horsetraders or Persian rug sellers in the Casbah, we don't want to haggle or negotiate the price each time we make a deal. Therefore, we have devised the rate card. Without one you can be lost at sea.

A rate card (see Figure 6.1) serves two major functions: it is a menu of the services you offer, and it lists the standardized hourly, daily or weekly rate you charge for those services. If you wish to design a rate card for your company, here are some specific steps to follow.

First, decide what standard services you are prepared to offer. The more you

**Figure 6.1: Rate Card for Independent Video Production Company**

## Remote Location Rate

### ¾" EFP:
$45.00 per hour — 1 hour minimum
20¢ per mile
$250.00 day rate — 8 hour day

### 35mm SLIDES:
$25.00 per hour — ½ hour minimum
20¢ per mile
$ 3.00 per slide — $2.00 per dupe

## Video Tape Editing Rate

### ¾" U-MAT SUITE:
$35.00 per hour — ½ hour minimum
$45.00 per hour with operator

### DUBBING:
$1.25 per minute plus tape
¾" to ¾"
¾" to VHS
¾" to BETA
VHS to ¾"
BETA to ¾"
5 Minute Minimum

## Video Tape Purchase

### ¾" VIDEO CASSETTES:
UCA  5 min. — $20.00
UCA 18 min. — $25.00
UCA 30 min. — $30.00
UCA 60 min. — $35.00

### ½" VIDEO CASSETTES:
VHS T30-120 — $20.00
BETA L250 - 750 — $20.00

## Legal Video Services

### DEPOSITIONS:
$200.00 plus 20¢ per mile

### DAY IN THE LIFE:
$250.00 plus 20¢ per mile

**TERMS OF BUSINESS:** Net 10 days. Accounts past due 30 days or more shall have a 1½% per month service charge applied.

**NOTE:** Short Notice, Unscheduled or Weekend Hours may be subject to overtime (1½) rates.

can provide on a continuing basis, the better are your chances of getting a fairly steady flow of business—and income.

If you have a modest post-production facility, with at least two tape decks and a system for enhancing and transferring audio, you may want to rent out that facility for offline editing, dubbing, etc. The addition of a Beta and a VHS VCR enables you to offer similar services to the rapidly expanding home and semiprofessional fields. A rental arm of your facility can bring in a steady flow of cash in many areas of the country. It is a bad idea, however, if you anticipate doing a great deal of your own production or can't afford the increase in equipment maintenance costs that will result from extra use.

Many producers started out as writers. If you are one of them, offer writing as one of your services. Many writer-producers make nearly as much money doing everything from technical manuals to newsletters and brochure copy as they do from producing.[1]

Once you've decided what services to offer, put a price tag on each of them. In order to be competitive in your marketplace, get rate cards from other producers and from your local television stations and see what their prices are. Local merchants who advertise on TV often employ local stations to do their productions. Advertising agencies with no in-house production capabilities also use TV stations. You want to take away some of that business, which means you will be less than popular with the management of the stations. Use some tact, therefore, in getting your hands on their rate cards.

Once you have an assortment of these cards from stations and other independent producers, compare their prices with what you had hoped to charge for your services. Two types of businesses charge the lowest rates: the large outfit that does enough volume to allow for reduced rates, or the one-person operation trying to undersell everybody else. You will no doubt find a comfortable position somewhere between the highest and the lowest figures quoted.

After calculating your real overhead costs (the very minimum you have to make in order to break even), add on a reasonable charge for your time and profit for the company. Never forget that the company is a separate entity from you and any employees you have. It must have its share of each project to stay alive.

There are no standards for profit margin and owner's salary. Some producers work regularly with 100% markups and others squeak by on 15% to 20%, relying on high volume. A more typical profit margin runs between 40% to 50% of the budget. You should calculate a minimum profit for your company by considering the real expenses of day-to-day operation (what it costs to keep your doors open), what your salary is going to be, and what financial cushion you need to continue operating between productions.

Use before-tax profits to sustain the business, purchase and maintain equipment, provide for expansion, etc. Put after-tax profits to work in interest-bearing savings accounts or other investments for which you should consult a financial advisor. A good rule-of-thumb is that businesses must grow at least 10% per year to survive the onslaught of inflation.

Determining your own salary is a matter of conscience. Many producers just starting out take no salary beyond the necessities of food, rent, transportation, etc. They pump all income into the business. This is neither fair nor wise. Suffering for the cause may lead to anger, frustration, resentment of employees and other traumas.

Pay yourself a fair salary in relation to the business and its employees. Remember that you will need a vacation and that your family, if you have one, should not have to suffer unduly because you have gone into business for yourself. As that business grows, you are entitled to the lion's share of the salary schedule since you have taken all the risks and, without you, there would be no company.

Although it is tempting to try to undercut the competition as a method of taking away business, use extreme caution if you decide to use this approach. Remember that a number of factors go into the decision of a corporate executive or small business owner to contract for a service—the cost of that service is sometimes one of the last considerations.

Businesspeople know that quality costs money. They are, for the most part (as you should be), very suspicious of anyone who promises high quality at an exceptionally low price. The terms are antipodal.

A note of caution here. Some states have laws to prohibit unfair competition, and impose a minimum profit margin on businesses. In Wisconsin, the minimum is 6%. Check with your state to find out if this law applies to you.

Remember, too, that if you are cast in the role of Cheap Sam, you may *never* be able to raise your rates. When negotiations with a client hinge on cost, it is always to your advantage to be able to come down a percentage point or two to clinch the deal. It is *never* wise to raise your price when negotiating a deal.

When you have decided on an appropriate rate structure—one that is fair and in line with your competition—have it typeset and printed. Include your logo and a brief promotional pitch if you choose. The card will become a fixed part of your presentation package. It can be adjusted annually to account for inflation, market fluctuations and so forth. Exercise due care when making it up, for it will become the basis for your tenure in this business.

## BILLING AND COLLECTING

It seems like a simple matter to charge somebody for your services, send them a bill and collect your money. For some reason—in this business especially—it frequently isn't this easy.

The product that we sell is more intangible than a car or a refrigerator. Once it's completed, it runs for a brief moment on a screen. Then, as far as clients are concerned, it is gone. Perhaps that is why it is sometimes difficult to wrestle payment from clients and why the single biggest complaint of small and medium-sized independent producers is slow-to-no collection of bills.

At the outset of a production, clients are usually excited, cooperative and enthusiastic. At this stage, they are quite friendly with the producer. In the best of all possible worlds, this beginning stage is the best time to collect the full budget.

Unfortunately, as in most other business transactions, the bill is not rendered until the product or service has been delivered. By that point, the client is often disinterested, disenchanted and sometimes suffering the post-production blues that are common to many producers as well. This odd syndrome is comparable to the post-partum blues of some new mothers. After all the anticipation, exhilaration and camaraderie of the production, suddenly it is gone with the birth of the finished product. Life seems mundane again.

Producers may have other productions in the offing and can cure these blues by getting back to work. Clients may have no such diversion. They may vent their frustration by inventing reasons not to pay for the production. Clients' blues may be further aggravated because they have no immediate feedback on the success of the project. They may begin to doubt the validity of their initial decision to spend a great deal of money on something so ephemeral as a video production.

Before a project begins, there is no sure way of determining whether clients will pay their bills on time or at all. However, the following sections offer independent producers some guidelines for correct, businesslike billing procedures and discuss what actions can be taken when you're confronted with clients who will not pay.

## The Down Payment

Most independent producers collect a substantial down payment in advance of each production. Usually the payment schedule is designed so that each of the three major phases of a project is paid for along the way. Once you have a complete budget approved by clients, tell them that payment would be appreciated as follows: one-third on script approval; one-third on completion of principal photography; and one-third on delivery of the completed tape.

Some producers divide the budget four ways, calling for one-fourth on approval of the proposal to develop the script, one-fourth on script approval, and so on. With either method, the idea is to refrain from dipping into your own pocket during a project. The final payment on delivery of the tape is normally the profit for the job. If, for some reason, you are not paid this final installment, you should at least break even on the deal.

## Keeping the Client Involved

During production and post-production, make clients feel as though they are part of the team. While you can't and shouldn't tolerate unnecessary interference from clients, you do want to cultivate a feeling of mutual respect. Put the dazzle of "making television" to your advantage. While the cameras are rolling and everything is in full swing, be sure to escort the clients onto the set (while keeping them out of the way of your director). If you have a wrap party at the end, as most producers do, be sure that the clients are there and that compliments and thanks for the job are passed around.

In post-production, when you have the offline ready to go online, invite the clients in to see some of the final stages of the process. You don't want clients breathing down your neck during the hard work and tough decision making. You don't want clients seeing all the blown takes and goofing off that frequently show up in offline. But letting them in at the finish builds good rapport.

## Drafting Professional Bills

Establish a standardized procedure for billing as part of your general accounting. Send itemized statements on a regular basis. Develop a form of your own or adapt one from the numerous printed forms available at office supply firms.

A professional billing statement is the hallmark of good business. It should have a printed logo and include at least the following:

- Description of the service rendered
- Date of completion of this service
- Contract reference number
- Total amount of the initial budget
- Amount already paid on account
- Balance due now
- Statement of terms of contract
- Statement of interest or service charges for unpaid balance. (This is normally 1½% per month after 30 days. Some states have usury laws that put limits on the amount of interest you may charge. Check the law in your state before adding interest charges.)

## Collection Problems

If the clients have not paid their bill after a reasonable period of time, attempt to work with them personally. Offer to meet and discuss the situation. If clients are having trouble with cash flow, offer to work out a monthly payment schedule.

Always present yourself as a rational, good-natured problem solver. Most people respond well to an honest attempt to communicate; threats and temper explosions never work.

If you still encounter resistance after about 90 days of trying to resolve the problem, you have two alternatives: you can go to court or hire a collection agency. Neither is pleasant and neither guarantees that you will ever get your money.

## Going to Court

You will, of course, want to consult an attorney before deciding whether litigation is a viable solution. For the most part, lawsuits for unpaid bills are tricky. You must be sure that you have something in writing—in legal terms this is called a "writing." This can be a letter of agreement or a contract. For very small amounts, a "writing" may not be required. In any case, you're on firmer ground if you have a signed contract.

In spite of what you've heard about verbal contracts, in most cases they aren't worth the breath it takes to make them. In court it will come down to your word against the client's word. Usually, that is not enough for you to win.

The other tenuous aspect of a lawsuit is that you will no doubt have to prove that through no fault of your own, a breach of contract has taken place. We'll detail this in the next section on contracts.

If you go to court, you have two options: you can either go to small claims court or file a more complex civil suit. The maximum awards for small claims court vary from state to state, but normally range between $500 to $1500. You do not need to hire an attorney in small claims, although both you and the defendant may do so. You can file the papers for the lawsuit at the Clerk of Courts office; a modest fee is charged.

If the defendant does not appear at the hearing, or otherwise raise an objection, it is possible for you to win the suit by default. If the defendant does show up, you will have to prove to the satisfaction of the court that the defendant owes you the amount claimed. If you prove this, the judge may award you a judgment for that amount.

Bear in mind that the defendant may file a countersuit against you for damages that could be in excess of what you sued for in the first place. The court will weigh both sides of the issue and make a determination. The appeals process for small claims courts varies widely. In some states, the decision of the judge is final. In other states, you or the defendant may make an appeal to a higher court within a specified amount of time. The best advice is to know your state laws before you enter into any action.

For dollar amounts in excess of the maximum for small claims, you must bring a more complex civil suit. Most courts in America have tremendous backlogs of pending cases. Filing a civil suit and waiting for the case to appear before a judge and jury can be an incredibly time-consuming process. In many states it can take from one to five years before your case is ever heard. You will also have to hire an attorney, whose fee can exceed the amount of money for which you are suing. If you win the case, it is sometimes possible for you to get a judgment for attorney fees in addition to the amount of the suit. This is not guaranteed in any case. Nor does winning the suit mean you can rely on getting any money in the end.

Receiving a judgment does not guarantee that the defendant will ever pay you. It simply means that the court agrees that you are owed the money. Collecting the money involves another set of legal maneuvers that may end up costing you more than the amount of the judgment.

Once you have filed a suit and your clients realize that you are serious about being paid, it may be best to negotiate out of court with them and settle for a sum lower than the amount actually owed. Defendants will often settle out of court rather than risk a judgment in the full amount or incur the expense of hiring an attorney and defending the suit. And once a lawsuit is looming, chances are good that clients will pay the amount agreed upon.

As with all legal matters, if you have any doubts or questions about proceeding, check with an attorney.

### Hiring a Collection Agency

When all other avenues have failed to get a debtor to pay up and you've decided that a lawsuit is too much bother to be worth it, you have two options. One is to give up and write the episode—and the loss—off to experience. The other is to turn the bill over to a collection agency.

These outfits generally work for a percentage of the total bill and you pay them only if they collect. Laws regulating collection agencies vary in each state, and there are restrictions on just how far an agency can go in its badgering of debtors.

Usually, credit agencies can send notices to the party threatening in vague terms some kind of "action" if the debt isn't paid. They can report bad debts to various credit agencies, which may influence the person's credit rating. If your client is concerned about such things, the collection agency just might be paid.

In general, though, the collection agency is the last resort. Do not count on it for much more than a final blast at the villian in your life, who has more than likely gotten away with not paying you. Move on quickly to more productive pursuits and cut your emotional losses by trying to forget about it.

## MAKING A CONTRACT

The contract is the most important document you will use in business. Not only is it your principal device for holding a client to the payment of your bills, but it is the foundation for all business you conduct with free-lancers and suppliers.

Producers frequently contract with technicians (such as editors, DPs, etc.), who are not regular employees of the producer's company. This is beneficial to the small producer who does not need to withhold taxes, pay for worker's compensation, unemployment insurance, etc., for such contracted work. Independent producers usually contract for rental equipment and supplies, too. Finally, if working in broadcast, you will also be dealing with distribution contracts for anything you sell to television, either network or syndicated.

Though there are elements common to all contracts, be advised that no contract is "standard." Contract law is complex, and you should always have an attorney draw up your contracts and interpret all but the most routine agreements presented to you by other parties.

In essence, a contract is simply the written form of an agreement between two or more parties before entering a project. The three basic elements of a contract are offer, acceptance and consideration. These three items are the heart and soul, the legally binding cement of any contract. Breaching, or not abiding by just one of them, will invalidate the contract and may open the breaching party to damaging civil action.

## The Offer

When drawing up a contract between you and a client in which you offer to produce a videotape, you must specify exactly what will go into this project. (Figure 6.2 is an example of a contract that could be used for the Jones Hotel project.) In your contract, specify the subject of the videotape, state that you will consult with the client, that you will write, or have written, a script for the client to approve and that you will make a videotape from this script to the best of your ability.

You must also state that the videotape will be of a certain length and format in its final delivered form, that it will take a certain amount of time to produce it and that it will cost a certain amount of money. The money will be paid to you in a certain prescribed manner according to a formula that you specify here in the contract.

In making the offer, be as specific as possible. Include such items as the kind of music (either prerecorded or original), whether the tape will be sync-sound or narrated voice-over, what actors will be in it, and so on. Remember that you will be held to providing everything you include in your offer.

Be very specific as to what the clients will be expected to provide. These are the conditions of the offer. Always provide for some sort of approval (e.g., if the script is approved and the project is an accurate reproduction on tape of the elements of the script, then the clients must accept the videotape). If you aren't careful in this area, you may encounter clients who refuse to pay because they don't like the product.

## The Acceptance

After you make your offer, it is imperative that the other party accept it with all the conditions that have been set forth. Typically, acceptance is at the end of the contract form with a statement that reads something like this:

**Figure 6.2: Sample Production Contract for a Video Project**

**MY OWN PRODUCTION COMPANY   123 1st Street   Anytown, U.S.A.01234**

**PRODUCTION CONTRACT**

1. The purpose of this document is to establish a contract between My Own Production Company, hereinafter referred to as PRODUCER and _____, hereinafter referred to as CLIENT for the production of a videotape program, hereinafter referred to as the PROJECT.

2. PRODUCER offers to produce the PROJECT for CLIENT in exchange for valuable consideration as specified in Paragraph 9 below.

3. DESCRIPTION OF THE PROJECT: The PROJECT is to be a ___ minute, color, sound videotape about the procedures used by the houskeeping staff of the CLIENT in the operation of CLIENT'S Hotel/Motel business. The purpose of the PROJECT is to provide CLIENT with an audio/visual training device for use in indoctrinating present and future housekeeping staff in their duties as housekeepers. No other use of the PROJECT is anticipated, nor have terms of consideration been agreed to for any but the use specified herein.

4. PRODUCER shall provide CLIENT a written script in a format standard to the video production industry prior to the principle photography of production. CLIENT shall have the right to approve said script for production or to stipulate changes thereto in order to make said script suitable for approval by CLIENT. Said approval shall be in writing with a signature block for CLIENT on the front page of the final script to wit: THIS SCRIPT IS APPROVED FOR PRODUCTION

5. PRODUCER shall, subsequent to script approval, exercise sole discretion in the hiring and firing of cast, crew and other production staff, exercising his best professional judgment in this selection to insure the highest possible standards of production for the specified budget.

6. CLIENT shall approve the written production budget submitted in advance by the PRODUCER. PRODUCER warrants that the production shall cost no more than the grand total listed on said budget without prior written consent of the CLIENT.

7. PRODUCER shall submit a shooting schedule to CLIENT for approval. Once said shooting schedule is approved in writing, any deviation or delay therefrom caused by CLIENT shall be considered unreasonable and any costs for said delay shall be charged to the production budget over and above the original grand total of said budget.

8. PRODUCER shall provide a finished version of the PROJECT to CLIENT no later than the ____day of _____, 19___ at the CLIENT'S place of business which is Address   City   State Zipcode.

**Figure 6.2: Sample Production Contract for a Video Project (Cont.)**

9. The grand total of the production budget for PROJECT approved by CLIENT is $_____. This total is payable as follows:
    [a] 25% at contract signing, receipt of which is hereby acknowledged by PRODUCER.
    [b] 25% upon written script approval.
    [c] 25% upon completion of principle photography.
    [d] 25% upon delivery to CLIENT of the finished videotape.

10. There are no other agreements or codicils relative to this PROJECT between either of the parties hereto either verbally or in writing and this document constitutes the entirety of the contract.

I have read, understood and agreed to each and every provision of this contract and with my signature, hereby certify and avow that I accept and agree to abide by them and that I am competent and legally qualfied to enter into such a contract.

Executed this _____day of _____, 19___ in the city of _____, County of _____, State of _____ by:


_____          _____
CLIENT                          PRODUCER



_____
WITNESS

I have read, understood and agreed to abide by all of the provisions of this contract on this_____day of_____, 19____ in the City of _____, State of _____.

Both parties to the contract sign in spaces provided beneath this acceptance. A witness to the signatures (preferably a Notary Public) is highly advisable. At least two copies of the contract must be signed in ink. One copy is for you, the other is given to the client. It is a good idea to make an extra copy to give to your attorney.

## Consideration

"Consideration" is the aspect of a contract most open to dispute, interpretation and misunderstanding. Think of consideration as both parties' inducement to enter into the contract. For a production contract, producers will get an agreed-upon sum of money for their work and clients will get a completed videotape. Most contracts contain wording like, "For X amount of dollars and *other valuable consideration,* the parties hereto agree," etc.

The "other valuable consideration" may be the privilege of doing business with a client with a big name, the opportunity to prove oneself worthy of further work, the enjoyment of associating with a celebrity, etc.

Nothing in the preceding section has been written or offered as legal advice. This section has been informational in nature only. For assistance with any contractual needs, consult an attorney in your state.

## BARTERING

Another standard practice in this unstandardized business is called "trade-out," and it involves the ancient concept of bartering your services. A few examples of bartering are given below.

You might, for instance, get a car dealer or manufacturer to lend you a car for a shoot, provided the vehicle is prominently featured and the supplier is acknowledged in the credits.

You might trade the production of a TV spot in exchange for services. For example, you shoot a commercial for Joe's Garage, and Joe services and repairs your car. You could also trade a screen credit line to a local cafe in exchange for free meals for the cast and crew during production. You might even shoot a scene at the cafe, making sure that the name of the place appears.

In trading out, let your imagination be your guide. It never hurts to ask; the worst that can happen is that the prospect will say no.

## SUMMARY

Our purpose, in part, has been to present you with an overview of information based on the practical, real-life experiences of dozens of independent video producers. A great amount of research has also gone into providing you with methods for keeping a continuous flow of information about this business coming your way.

Another purpose, stated at the outset, has been to give you a comprehensive vision of what a producer is, what a producer does and how a producer sets up a business and runs it. What may not be apparent is the love.

Most producers will confide, quite frankly, that going into independent production is a nutty thing to do. The purely rational mind would see at the outset that it makes more sense to work at a job where the checks are regular and somebody else has the headaches. The desire to be an independent producer is not rational.

In the end, the producer is largely an unsung hero. Remember this: If you are the producer and the project fails, it will be your fault. If it succeeds, it will have 100 mothers and fathers—none of them you.

Accolades will be heaped on the actors, the director and the DP, all of whom deserve them, of course. But none are likely to come your way. This odd truth of our business is something you simply will have to learn to live with, drawing whatever satisfaction you can from the awareness of your own accomplishments as the prime mover in the piece. Producing is, by and large, a thankless job.

And that's where the love comes in. You have to love doing it.

Of all the producers who contributed their time and combined wisdom to this book, of all the producers who ever sweated a project through to the end, most often misunderstood and unappreciated, not a single one can think of anything else they would rather be doing with their lives. If you are one of us, good luck...and welcome aboard!

## NOTE

1. For an annual listing of free-lance markets for writers, see *Writer's Market* (Cincinnati, OH: Writer's Digest Books, 1985).

# Appendix A:
# Directory of Guilds and Unions

## Actors and Artists

American Federation of Television and
    Radio Artists (AFTRA)
1717 No. Highland Ave.
Los Angeles, CA 90028
(213) 461-8111

Screen Actors Guild (SAG)
7750 Sunset Blvd.
Los Angeles, CA 90046
(213) 876-3030

Screen Extras Guild (SEG)
3629 Cahuenga Blvd.
Los Angeles, CA 90068
(213) 851-4301

## Technicians

International Alliance of Theatrical
    Stage Employees (IATSE)
14724 Ventura Blvd.
Sherman Oaks, CA 91403
(818) 905-8999

National Association of Broadcast
    Employees and Technicians
    (NABET)
333 No. Glenoaks Blvd.
Suite 640
Burbank, CA 91502
(818) 846-0490

## Directors

Directors Guild of America
7950 Sunset Blvd.
Los Angeles, CA 90046
(213) 656-1220

## Writers

Writers Guild of America, West
    (WGA)
8955 Beverly Blvd.
Los Angeles, CA 90048
(213) 550-1000

## Transportation

Studio Transportation Drivers Local
    399, International Brotherhood of
    Teamsters
4747 Vineland Ave.
Suite E
North Hollywood, CA 91602
(818) 985-7374

## Sound

Broadcast TV Recording Engineers
    International Brotherhood of
    Electrical Workers (IBEW)
3518 Cahuenga Blvd. W.
Suite 307
Los Angeles, CA 90068
(213) 851-5515

International Sound Technicians,
    Cinetechnicians and TV Engineers
    Local 695, IATSE
11331 Ventura Blvd.
Suite 201
Studio City, CA 91604
(818) 985-9204

Sound Construction Installation &
    Maintenance Technicians Local 40
    IBEW
5643 Vineland Ave.
North Hollywood, CA 91601
(818) 877-1171

## Musicians and Composers

American Guild of Authors &
    Composers/The Songwriters Guild
6430 Sunset Blvd.
Suite 1113
Los Angeles, CA 90028
(213) 462-1108

American Guild of Musical Artists
    (AGMA)
12650 Riverside Dr.
Suite 205
North Hollywood, CA 91607
(818) 877-0683

Musicians Union Local 47 American
    Federation of Musicians/AFL-CIO
817 No. Vine St.
Los Angeles, CA 90038
(213) 462-2161

## Talent Agents

Association of Talent Agents
9255 Sunset Blvd.
Suite 930
Los Angeles, CA 90069
(213) 274-0628

## Wardrobe and Makeup

Association of Film Craftsmen Local
    531 NABET
1800 No. Argyle St.
Suite 501
Los Angeles, CA 90028

Costume Designers Guild Local 892
    IATSE
14724 Ventura Blvd.
Penthouse
Sherman Oaks, CA 91403
(818) 905-1557

Motion Picture Costumers Local 705
    IATSE
1427 No. La Brea Ave.
Los Angeles, CA 90028
(213) 851-0220

The location, addresses and phone numbers of regional and East Coast offices of
all of the above guilds and unions are available by inquiring to the offices listed here.

# Appendix B:
# Selected Distribution Companies

The following is a listing of several major distribution sources of nontheatrical film and video.

## SELECTED BROADCAST TELEVISION DISTRIBUTORS

Alan Landsburg Productions
1181 W. Olympic Blvd.
Los Angeles, CA 90064
(213) 208-2111

Arista Films, Inc.
16027 Ventura Blvd.
Encino, CA 91436
(818) 907-7660

Bob Clampett Productions, Inc.
729 Seward St.
Los Angeles, CA 90038
(213) 466-0264

C & C Syndication
4501 Greengate Ct.
Westlake Village, CA 91361
(818) 889-1064

Century Distributors, Inc.
16153 Cohasset St.
Van Nuys, CA 91406
(818) 781-0177

Children's Media Productions
PO Box 40400
Pasadena, CA 91104

Columbia Pictures
Columbia Plaza
Burbank, CA 91505
(818) 954-6000

Dan Curtis Productions, Inc.
c/o ABC Circle
9911 W. Pico Blvd.
Los Angeles, CA 90035
(213) 557-6910

Dick Clark Productions
3003 W. Olive Ave.
Burbank, CA 91505
(818) 841-3003

Four Star Entertainment Corp.
19770 Bahama St.
Northridge, CA 91324
(818) 709-1122

Gold-Key Entertainment
931 No. Cole St.
Los Angeles, CA 90038
(213) 469-2102

Home Box Office
2049 Century Park E.
Suite 4170
Los Angeles, CA 90067
(213) 201-9250

Jerry Dexter Program Syndication
139 S. Beverly Dr.
Beverly Hills, CA 90212
(213) 278-9510

Lorimar TV Distribution
3970 Overland Ave.
Culver City, CA 90230
(213) 202-4204

MCA-TV
100 Universal City Plaza
Universal City, CA 91608
(818) 985-4321

Metromedia Producer Corp.
5746 Sunset Blvd.
Los Angeles, CA 90028
(213) 462-7111

MGM Entertainment Co. TV
10202 W. Washington Blvd.
Culver City, CA 90230
(213) 558-5000

NBC Network Sales
3000 Alameda Ave.
Burbank, CA 91523
(818) 840-4444

Orion Pictures Corp.
1875 Century Park E.
Los Angeles, CA 90067
(213) 557-8700

Paramount Pictures, Inc.
5555 Melrose Ave.
Hollywood, CA 90038
(213) 468-5000

Samuel Goldwyn Television
1041 No. Formosa Ave.
Los Angeles, CA 90046
(213) 650-2400

Twentieth Century Fox
10201 W. Pico Blvd.
Los Angeles, CA 90035
(213) 277-2211

Viacom Enterprises
10900 Wilshire Blvd.
7th Floor
Los Angeles, CA 90024
(213) 208-2700

The Vidtronics Co.
855 No. Cahuenga Blvd.
Los Angeles, CA 90038
(213) 856-8200

Warner Brothers TV Distribution
4000 Warner Blvd.
Burbank, CA 91522
(818) 843-6000

Wrather Corp.
270 No. Canon Dr.
Beverly Hills, CA 90210
(213) 278-8521

## SELECTED NONBROADCAST GENERAL DISTRIBUTORS

Cinevid, Inc.
720 Seward St.
Hollywood, CA 90038
(213) 465-9076

Cori Films International
2049 Century Park E.
Suite 1200
Los Angeles, CA 90067
(213) 557-0173

Creative Video Systems
7920 Alabama Ave.
Canoga Park, CA 91304
(818) 888-3040

Encyclopaedia Britannica Educational
   Corp.
425 No. Michigan Ave.
10th Floor
Chicago, IL 60610
(312) 321-7410

Films, Inc.
5625 Hollywood Blvd.
Los Angeles, CA 90028
(213) 466-5481
or
733 Green Bay Rd.
Wilmette, IL 60091
(312) 256-6600 through 6611

Intercontinental Releasing Corp.
9000 Sunset Blvd.
Suite 1000
Los Angeles, CA 90069
(213) 550-8710

MGM Home Entertainment Group
5890 W. Jefferson Blvd.
Los Angeles, CA 90016
(213) 838-7343 and (800) 223-0933

Modern Talking Picture Service, Inc.
6735 San Fernando Rd.
Glendale, CA 90049
(818) 240-0519

Pyramid Films
2801 Colorado Ave.
Santa Monica, CA 90404
(213) 828-7577

Roundtable Film & Video
113 No. San Vicente Blvd.
Beverly Hills, CA 90211
(213) 657-1402

Telepictures Corp.
15303 Ventura Blvd.
Suite 1201
Sherman Oaks, CA 91403
(818) 986-3600

Westcom Productions
9000 Sunset Blvd.
Suite 415
Los Angeles, CA 90069
(213) 278-0112

For a more complete list of nonbroadcast distributors, see *The Video Register,* by the editors at Knowledge Industry Publications, Inc. (White Plains, NY: Knowledge Industry Publications, Inc., 1985).

# Appendix C:
# Selected Video Trade Publications

Keeping abreast of the rapidly changing technologies and procedures of the video business is essential to any producer. In addition to attending trade shows and conferences, a good way to keep in contact is through trade publications.

There are many periodicals that cover the video business. They run the gamut from looseleaf notes to slick weekly or monthly magazines. Many are available free of charge to bona fide independent producers; others are available through paid subscription. Few are available on newsstands or at bookstores.

Following is a list of some major publications that independent producers may find useful:

*Audio-Visual Communications*
475 Park Ave. So.
New York, NY 10016

*Audio-Visual Communications* is a monthly magazine of about 100 pages per issue that features informative articles on new equipment, techniques and tips from professionals.

*BM/E*
*Broadcast Management Engineering*
295 Madison Ave.
New York, NY 10017

*BM/E* is a monthly magazine of about 100 pages available free to anyone in charge of purchasing broadcast equipment or for station operations. If you're a technology buff, this one is for you. Not for the casual reader, *BM/E* can be of great value to the producer who wants to keep abreast of the very latest technology. One major advantage of this publication is that subscribers receive an annual equipment and services yearbook.

*Cable Marketing*
352 Park Ave. So.
New York, NY 10010

*Cable Marketing* is a slick monthly tabloid free to operators of CATV systems and their corporate headquarters. Brisk and lively, it keeps readers informed of all the latest developments and problems in the cable industry. *Cable Marketing* is a primary source for any producer wishing to know about, and sell, products for cable systems.

*CURRENT*
Box 53358
Washington, DC 20009

*Current* is a twice-monthly newspaper of about 10 pages that deals specifically with public broadcasting news, policies and procedures. It is a good source of information on programming possibilities, the latest legislation affecting Public Broadcasting Service (PBS) and National Public Radio (NPR), and many other issues of concern to producers of public broadcasting shows.

*Electronic Media*
740 Rush St.
Chicago, Il 60611

*Electronic Media* is a weekly tabloid magazine of about 100 pages and a major news organ for television. Coverage of major network and syndicated TV doings is bright, informed and very readable. For any producer who wishes to break into the "majors," *Electronic Media* is a must.

*Industrial Photography*
475 Park Ave. So.
New York, NY 10016

The oldest magazine devoted exclusively to industrial audiovisual producers, this monthly is a very fine publication for the all-around AV producer. Subscribers receive the annual Goldbook with complete listings of equipment manufacturers, distributors and rental houses.

*Shooting Commercials*
PTN Publishing Corp.
101 Crossways Park W.
Woodbury, NY 11797

A monthly tabloid magazine of about 35 pages, *Shooting Commercials* is mandatory reading for anyone who wants to produce television commercials. It is newsy and highly readable, and keeps readers up to date on the doings of other producers, ad agencies, etc. Articles give tips, discuss new ideas and technologies and tell you what the big advertisers are doing and planning.

*Television Broadcast Communications*
Globecom Publishing, Ltd.
4121 W. 83rd St.
Suite 265
Prairie Village, KS 66208

A monthly of about 150 pages, *TBC* is largely aimed at broadcast managers and technicians. It has fine articles on technological developments, news, facility design, etc. Producers may find occasional insight into broadcast trends and thinking.

*Video Manager*
Knowledge Industry Publications, Inc.
701 Westchester Ave.
White Plains, NY 10604

A tabloid magazine of about 30 pages, VM is published monthly except for a combined July/August issue. It is written especially for managers of video networks in business and industry, cable, medicine, government and education, as well as for production houses, dealers and consultants. Articles discuss theory and techniques, industry events and hardware development. The magazine also provides both hard news and feature stories about the industry.

*Video Systems*
PO Box 12901
Overland Park, KS 66212-9981

Subtitled "the how-to magazine for video professionals," *Video Systems* is a slick, monthly magazine format publication averaging 70 pages. Twice a year, in May and November, it includes a "VideoIndex" section that lists the names and addresses of video production companies. The magazine has many full-page ads representing major manufacturers of goods and services with a reader service card for more information. *Video Systems* is an excellent, lively, fact-filled magazine—a must for any serious producer.

*Video Trade News*
Tepfer Publishing Co.
51 Sugar Hollow Rd.
Danbury, CT 06810

*VTN* is a tabloid monthly newspaper averaging 12 pages. It gives general news about people, companies, products and the political scene in Washington as it may affect the industry. The advertising and classified sections include "Help Wanted" columns. It is a very comprehensive small newspaper.

*Video Week*
Television Digest, Inc.
1836 Jeffeson Place, NW
Washington, DC 20036

*Video Week* bills itself as, "Devoted to the business of program sales and distribution for videocassettes, disc, pay TV and allied news media." It is a weekly, eight-page newsletter format, drilled with three holes to fit in a binder for easy reference. The tone is breezy with succinct summaries that presume the reader has a background in the business. Since the subscription price is rather high, you should have a look at the free trial before considering a subscription.

# Appendix D:
# Advice from Hollywood Producers

A good way to determine your own strategy for doing business is to seek advice from successful producers and to study the examples set by others who have preceded you.

Following is advice from top Hollywood producers Saul Bass, Henry Winkler and Saul Turtletaub. These excerpts were taken from taped interviews conducted in Hollywood, CA, in 1984 as part of an off-campus course taught by the author at the University of Wisconsin-Oshkosh. These producers' comments, especially those of Turtletaub, apply directly to formulating a game plan for landing a major account.

## SAUL BASS

Saul Bass is a man of many talents. He is a graphic artist, a designer, a director and a producer. He is most famous for directing the shower scene in Alfred Hitchcock's *Psycho* and the award-winning animated short, *Why Man Creates*. This is how he defines a producer and the method necessary to achieve a producer's goals:

> A producer can be the money man. He can be the guy who puts the deal together; or he can be the guy who does the nitty-gritty, who tends to move into the production area as the director's point of reference. He may be the guy who's developed the whole property and brings in the script, which is a damned important creative contribution, and who later helps with the editing of the piece.

Bass has a set of rules that he follows into any new project, from starting a design to nurturing an account.

> You have to have a procedure, an approach to a project or to your business. You have to identify the objectives of the exercise. And in

order to do that, you have to understand what has to be communicated.

If you want to do something for a company, you have to look at that company. You have to understand who they are, what they are, what their objectives are and where they are going or want to go. And therefore, you develop what amounts to an informal *sense;* a set of criteria or objectives so that anything you do has to fulfill those.

## HENRY WINKLER

Henry Winkler is perhaps best known for his role as ''The Fonz'' on the television series, ''Happy Days.'' He is lesser known as a successful independent producer with his own company, Fair Dinkum Productions. Here is what he has to say about producers in business for themselves:

> If you're going to be in this business, you've got to be tenacious. You have to believe in yourself and you have to know ultimately what you want. Everybody is going to tell you what you want is no good. And you have to understand that rejection has nothing to do with your body odor or perfume or the way you dress yourself. It has nothing to do with you personally, but with another person's fear of holding onto his or her job. It has to do with a political scam that you might not even know about going on in an organization.
>
> Remember that the people you're dealing with are insecure. You must always make them feel as if you are going to do them a favor. And, you have to really keep your mind clear about what it is you want and what it is you will do to get what you want.

## SAUL TURTLETAUB

Saul Turtletaub writes and produces situation comedies for television. When it comes to selling them, the comedy is very serious. Turtletaub's formula for approaching a network with a new series applies just as well to your selling an idea to General Motors or John Deere.

> The steps to follow are simple. First, you have everything set in your mind completely. You know exactly what it is you want to sell and you can say it succinctly. If you can't tell me what the story is in five minutes then you're sunk. Ninety percent of the sale depends on your ability to communicate verbally and to 'hook' the guy you're selling right away. Being a good story-teller is as important as being a good writer or producer. After this initial discussion, the steps follow in order:
>
> 1. Present a pilot script for which you're paid a reasonable sum.
> 2. Rewrite it with the client's input.
> 3. Read and test the project and get either a go-ahead or a go-away.
> 4. Move on with the shooting or get to work on something else immediately.

# Appendix E: Sample Proposals

**Jackson·Walsh**
**&** Associates, Inc.

THE GREAT AMERICAN DREAM MACHINES

Proposal for an advertising account contract
presented to:

Mr. Clyde Fessler
AMF/Harley-Davidson
Harley-Davidson Motor Co., Inc.
Milwaukee, Wisconsin

Account Executive: Victor Feathers

3923 W. 6th. Street Suite 216
Los Angeles California 90020
213 • 387-3231

<u>A WORD ABOUT US</u>

Creative.

That's the word. But, there is a LOT of philosophy behind the word; behind us.

We're an agency which believes in America. In American products. The American Dream is a reality for us, as well as for the clients we represent. We want AMF/Harley-Davidson on our team. In this proposal we will show you why.

We selected Harley-Davidson for one <u>major</u> reason.  You are America's motorcycle company. We have watched the Japanese machines, using American marketing and advertising practices, slowly erode your business. That business, according to your annual report to the stockholders was down 17% last year alone, primarily because your advertising was outstripped by Honda, Kawasaki and Yamaha.

We know that Harley-Davidson makes wonderful machines. Your powerful V-Twin engine, your workmanship, the legendary durability of the Harley-Davidson and the mile-after-mile comfort of your superbikes are all objects of pride for your company. AND...your products are made in America. So, why are your sales down?

Turn on a television set tonight and find out.

..."We <u>know</u>  why you ride"- Kawasaki

...a policeman giving the thumbs up sign on- a Honda.

Turn on a radio and listen to...

..."Kawasaki lets the good times roll"...

And we'll BET you can hum that tune JUST from reading that line of copy. We will have America humming the Harley-Davidson song. We can sell Harley-Davidson's. And THAT is all we need to say about us.

2.

<u>ON THE POSITIVE SIDE</u>
Sales!

That's what ANY agency promises to deliver. So do we. The
difference is that our fresh approach does it better, as the en-
closed sample campaign indicates. On page 4 we have some other
ideas which go outside traditional channels to sell the product.

Here we'd like to deal with some traditional specifics.
You spent nearly 60% of your advertising budget last year on print.
We recommend reversing the media mix with a heavy emphasis on broadcast media. Your current approach to large display ads in the motorcycle trade magazines misses a large segment of potential buyers.
An exciting radio and television mix will correct this deficiency.
Kawasaki spent 52% of its budget on television, for example, with
a resulting 210 million dollars in sales. Their magazine figures
were only 25% of budget, with newspapers getting 2%, radio 3% and
point-of-purchase only 9%. We would recommend for Harley-Davidson
a heavier concentration on radio which reaches over 98% of the
American population every day at very low cost per thousand figures.
Radio also offers very good specific demographic targets.

Our campaign is designed to reach the HEART first, then the
head. This also reverses your current campaign objective. Harley-
Davidson engineering IS superb. We doubt that the AVERAGE potential
buyer <u>cares</u>. What he wants is fantasy gratification.

Our initial campaign is a strong EMOTIONAL appeal.
<u>HARLEY-DAVIDSON- THE GREAT AMERICAN DREAM MACHINES</u>!  Encapsulated in
that slogan is a wealth of fantasy-building:

...patriotism...the mythos of obtainable wealth a la Horace Greely...

...prestige...a sense of belonging...power...national pride.

3.

We also move toward the nostalgia value of Harley-Davidson motorcycles with our banner lead:

<u>Remember the feeling you used to get? Harley-Davidson gives it back</u>!

There is strong appeal in evoking the sense memory of things like...

...your first pair of summer tennis shoes

...your first bicycle

...your first love affair

and so on. By linking these feelings to Harley-Davidson Motorcycles we have overcome a lot of the negative attitudes which many people still have toward motorcycles. To reach the guy in the suburbs who could become a Harley-Davidson buyer, we have many things to do away with in his mind:

...Fear of the machine, probably instilled by his parents with comments like, "They oughtta outlaw those things".

...Fear of the image of Hell's Angels; especially critical to Harley-Davidson Motorcycles.

...Fear of policemen; unfortunately still strongly identified with Harley-Davidson.

...Fear of being DIFFERENT! Astonishingly, recent research indicates that most Americans, in spite of what they profess, do NOT want to be different from their neighbors. We suspect this is the <u>main</u> reason one encounters so many "nice people" on HONDAs!

Paradoxically, these same fears can become the strongest POSITIVE factors in the decision to buy a motorcycle; especially if the fears are manipulated into fantasies.

On the next page we explore some concrete programs to describe this rather complex notion of human motivation.

4.

We are not <u>just</u> after the buyer who is ALREADY committed to
motorcycling. We want the Honda rider to switch to Harley, but,
more importantly, we want the first time buyer to "Go Harley" from
the start. Here are specific programs to overcome the fears we talked
about on the previous page.

1. <u>THE HARLEY-DAVIDSON GOOD GUY CLINIC</u>

Part of the fear of the machine is not understanding it. Our
dealers will hold evening clinics in motorcycling. Here, the current
or the prospective bike owner will take lessons in handling a motorcycle, riding in dirt, on the road and general safety and control
aspects important to the rider. For the expert rider we offer classes
in racing and stunt techniques under the close supervision of a professional rider. In short, the clinic is for everyone from rank
novice to road racer.

The second part of the clinic instructs riders in owner maintenance. This is in many ways, the scariest part of the game for
the suburban novice. He doesn't want to look like a fool when he
takes his girlfriend for a ride and the damn machine quits running.
Some basic instruction will eliminate his fear forever and make a
HERO of our Harley-Davidson dealer!

The clinic is open to anyone in the community, whether or not
they own a motorcycle, whether or not they own a HARLEY. And THAT
is WHY it's the "Good Guy" clinic. What we SELL is that Harley-
Davidson is interested in safer and more trouble free <u>motorcycling</u>,
period. AND...we're the only ones to be so public spirited. Bring
us your Honda...your tired Kawasaki and we'll show you how to take
care of it...while we sell you a Harley!

People who pass the course get a nifty patch for their jackets
and a decal for their bike which reads:
CERTIFIED HARLEY-DAVIDSON RIDER/MAINTENANCE SPECIALIST.

2. <u>THE HARLEY-DAVIDSON AMERICAN DREAM TOUR</u>

If you buy your Harley-Davidson during the month of July, we'll
FLY you to Milwaukee to pick it up and join our American Heritage Tour.
Our caravan with a maintenance and comfort vehicle in the lead, will
take you and your passenger through the Land of Lincoln to the his-
toric battlefield  at Tippecanoe. Our trail is over quiet country
roads...etc.

The idea here is very simple. The guy buys a bike to be "free",
but is scared to venture off on his own. So, we give him the dream
trip on his bike in perfect safety and in the company of other
riders who are EXACTLY like him. This idea made Wally Byam the King
of Caravans with his Airstream trailers. It satisfies the same urge
to merge with the great outdoors which fills every campsite in America
with motorhomes and campers. Our tour follows a simple route through
the Midwest with organized campsites at the end of days' short ride.
For an additional fee, we arrange shipment home of the bike at the
end of the tour and the customer gets a jacket patch and decal which
reads:
OFFICIAL HARLEY-DAVIDSON GREAT AMERICAN DREAM TOUR.

5.

Our Dream  Tour also takes care of another objection raised when a married man wants to buy a bike. Mom wonders how she's going to fit in. On our couples tour, she'll have the time of her life, and we can end up selling her a bike as well!

THE HARLEY-DAVIDSON SAFETY SCHOOL

A giant Harley-Davidson Number 1 van pulls up to the school yard. Out of it rolls a fleet of Harley-Davidson motorcycles with instructor/trick riders. We stage a one or two day demonstration for driver education classes, putting the high school kids in the saddle under close supervision. We have the local policeman there with his Harley Police Special, too.

Inside the van we have audio-visual presentations. Exciting films show our motorcycles in action: racing footage, Evel Knieval, motocross riding, trials competition, and plain old fashioned sport riding through the beautiful scenery of Door County.

The idea we're selling here is safety and interest in the true sport of motorcycling. Part of our Great American Dream is that the young rider will learn respect for the machines and will ride safely to live and enjoy the dream. This is another area where we can capture public attention as the "concerned motor-cycle company". To continue an on-going program of safety, we include our local Harley-Davidson dealer in the school's driver education program.

These are a few of the unique ideas which we propose as part of our service as your advertising agency. We know that in America, Harley-Davidson should be, can be and WILL be, Number One.

Since we are in Hollywood, in constant touch with the celeb-rities, the studios and the television production firms, and such much of our expertise lies in these areas, we can also promote Harley-Davidson motorcycles in a variety of ways aside from the conventional time-buy.

We invite you and whomever of your staff you feel would be appropriate, to come to Hollywood at our expense for two or three days. We want to prove that we can deliver and to show you how.

SAMPLE SPEC SPOT COPY

**Harley-Davidson Motorcycle Television Spot - 60**

Writer: Bob Jacobs        Account Exec: Victor Feathers

| VIDEO | AUDIO |
|---|---|

---

**1. LS ROAD**
A billowing American flag
SUPERED over. The road is
a twisting rural one lined
with trees in Autumn flame.
A rider, neatly dressed in
designer leathers comes toward us on a SPORTSTER. A
pretty lady is riding with
him, enjoying the ride.

dissolve to

**2. CU ANOTHER RIDER**
A good looking young exec
type. PULL BACK to reveal
another sylvan scene on the
California Coast. He and a
young, beautiful woman are
on a SUPERGLIDE near Big
Sur. Ocean sparkles B.G.

dissolve to

**3. MS ANOTHER RIDER**
He is on a full-dress
ELECTRAGLIDE rolling down
a rural road with red
barns and silos: American
"heartland". We FOLLOW.

**4. MONTAGE**
Quick cuts of a Motocross,
a young woman on an SX250,
a boy on an SX175 fording
a river, and a couple, typ-
ifying young urbans, setting up camp in the woods, 2
SS250's in FOREGROUND.

**5. CU HAPPY RIDER**
We pull back quickly in an
aerial shot and see 100
riders, all on various model Harley's in a procession
down a rural highway. These
are SUPERED over the Harley
Number 1 logo.

Fade to black.

---

(MUSIC UP- "Harley Song")
LYRIC
FREEDOM RIDES THE AMERICAN
ROAD, ALONG WITH YOU AND ME...

HARLEY PUTS YOU IN THE SCENE
FROM SEA TO SHINING SEA!...

ON BYWAYS OF THIS SPRAWLING
LAND,
PAST RIVER, LAKE AND STREAM...
HARLEY-DAVIDSON TURNS YOU ON
TO FEEL THE AMERICAN DREAM!
(Instrumental continues under)

NARRATOR

HARLEY-DAVIDSON IS AMERICA'S
MOTORCYCLE COMPANY. PART OF
THE AMERICAN DREAM SINCE 1903.
GETTING BETTER. PERFECTING OUR
MACHINES FOR AMERICAN ROADS,
AMERICAN RIDERS, AMERICAN
DREAMERS. RIDE THE HARLEY-
DAVIDSON DREAM MACHINE. YOU
WON'T WANT TO WAKE UP!

(Song continues)

COME JOIN THE HARLEY-DAVIDSON
CLUB AND FEEL THE AMERICAN
DREAM!

NARRATOR

HARLEY-DAVIDSON MOTORCYCLES-
THE GREAT AMERICAN DREAM MACH-
INES.

(FX- MOTOR SYNTHESIZED FADING)

# Jackson·Walsh
## & Associates, Inc.

COPY FOR HARLEY-DAVIDSON MOTORCYCLE PRINT AD     0002-77

(Banner-) OUR NUMBER 1 IS THERE FOR A REASON!

Actually it's there for several reasons.

We win races. LOTS of them.

We set and KEEP land speed records.

We set and KEEP jumping records with EVEL KNIEVAL.

(Ask HIM why we're number 1!)

But those aren't the IMPORTANT reasons.

YOU are.

You're an American rider. We're America's motorcycle company.

Harley-Davidson's have been built to be Number 1 with you since 1903.

Our V-Twin engine is the strongest one in town. Take it from L.A. to
New York sometime and see for yourself.

We EAT mountains, deserts, the great plains. Then we give you carefree
city street riding for dessert.

Our Number 1 is YOURS. Because  you're part of the American Dream.

HARLEY-DAVIDSON MOTORCYCLES- The Great American Dream Machines.

# Jackson·Walsh
## & Associates, Inc.

COPY FOR HARLEY-DAVIDSON MOTORCYCLE PRINT AD      0001-77

(Banner-) REMEMBER THAT FEELING YOU <u>USED</u> TO GET?...

...in your first new pair of tennis shoes for your tenth summer?

...on your very first Christmas bicycle?

...in your FIRST car?

...on your FIRST date?

HARLEY-DAVIDSON REMEMBERS THAT FEELING, TOO...

AND GIVES IT BACK TO YOU!

Harley-Davidson is AMERICA'S motorcycle company. We've been

part of the American dream since 1903. Getting better.

<u>Perfecting</u>  our machines.

Making them stronger. More durable. More FUN!

Our bikes are made for <u>American</u> riders. For <u>American</u> roads.

For American <u>dreamers</u>.

Our Number 1 is there for a reason. We've EARNED it, year after year,

in the toughest competition in the world- <u>American</u> racing!

You can find the "nicest people" anywhere. You can find YOURSELF on

a HARLEY-DAVIDSON.

HARLEY-DAVIDSON MOTORCYCLES- THE GREAT AMERICAN DREAM MACHINES.

THE MECHANISMS OF JOY

Proposal for a Recruiting Film for

Product Engineering Group
Mercury Marine

Presented to:

Neil A. Newman
Director of Product Engineering

by

Bob  Jacobs, Ph.D
October 15, 1985

**Where Ideas Grow**

1

## SECTION ONE

## INTRODUCTION

In our past couple of meetings, musings and meanderings around the world of Product Engineering at Mercury Marine, we have explored a problem which you have identified; a problem which lends itself to a unique and dynamic solution in the form of a quality film presentation. First, let's define and detail the problem. Then we will propose the format for the solution and the reasons why The Film Farm should be the choice to execute that solution.

You need to recruit high calibre design engineers to fill both present and projected vacancies in Product Engineering in the Fox Valley. The target candidate for these positions is a man or woman between the ages of 25 and 40, usually with a degree and/or extensive experience in mechanical engineering. The primary candidate would be someone already working in the field at places such as 3M, Texas Instruments, Ford, GM, Chrysler, Boeing Aircraft and  so on. The secondary candidate would be an undergraduate or graduate student at a major university. The salary range and fringe benefit package in this specialty field offered by Mercury Marine is competitive with other major corporations.

In spite of this apparent equity in the field, conventional methods of recruiting candidates for your positions have encountered obstacles, severely limiting your choice of the best people available. In our discussions, we have conjectured at several of the possible reasons why gifted young  engineers may

2

be reluctant to apply for work at Mercury Marine. These include all of the following, not necessarily in prioritized order:

[1] The perception by many people in other states that Oshkosh and Fond du Lac, Wisconsin are cultural backwaters is a significant consideration. The very fact that people in the industry who work in New York and Los Angeles say, "You must be from Oshkosh", when they mean, "You're a real hick" is indicative of the mentality in question.

[2] Another impression of our part of the country is difficult to refute. Winter is long, bitter, costly in terms of home heating, and frightening to many folks who are used to "the sun belt" or the beaches of Southern California.

[3] The recently highly publicized fact that Wisconsin is consistently one of the highest taxed states in the country, especially in the areas of personal income and property taxes, is a major deterrent to many American white-collar workers, especially to those in the salary bracket under consideration.

[4] There is a conception among the technical elite that boat motors are prosaic and that there is little if any challenge or excitement left in the field. Under the intense publicity blaze of recent space technology, many engineers feel that aerospace is the the cutting "edge of the envelope"; the place where the "action", and therefore **they** ought to be.

[5] Wisconsin itself, when it is thought of at all, is regarded by many otherwise intelligent people as an unexciting land of dairy farms, corn fields, beer-swilling, pot-bellied foundry workers and, even at this late period of political

3

history, the arch-conservative, closed-minded culture which has only produced Senator Joseph McCarthy and polka dancing.

Those of us who live here certainly recognize finely-ground kernels of whole grain truth in the foundations for such ideas about our state and the industry. The facts are that for nearly six months each year our trees are leafless, that from December through March the wind blows and the snow flows, that heating costs **can** eat up a large chunk of the family budget for the unwise and unwary, that one aspect of our society has a hard-drinking, hard-laughing, "blue-collar", four-wheel drive, Wednesday night bowling mentality, and that personal taxation is, indeed, a heavy burden for Wisconsinites to bear. And, if this were the **entire** story and 10 horsepower trolling motors for bass fishermen were the only challenge for a Mercury Marine engineer to contemplate, we would  all throw up our hands at **ever** luring a rational man or woman to such a terrible fate, pack up our bags and move **anywhere** else on the planet!

But, we know better.

We live here.

We work here.

We know that any whole picture is not represented by a few pixels therefrom!

And obviously, there is **something** which keeps us here. In fact, there are **many** things which keep us here and which, when presented properly to our prospective candidates, will encourage them to want to join us.

In the next few pages we will explore some concepts for

4

presenting the whole picture and the perfect format for doing it.

## THE MEDIUM IS THE MESSAGE

We have agreed that the format for this highly directional appeal should be a departure from the norm; a dramatic, motion-packed, emotionally stimulating "movie". There are two methods for recording whatever story we end up choosing. One is video. The other is motion picture film.

Before coming to our recommendations for the recording medium (film or video), we'd like to recap the justification for choosing this radical departure from more standard recruiting techniques.

First, it is impractical to expect that we can overcome the already indentified mindset of our candidate in a conventional print ad, or job announcement in trade/tech periodicals. There is, as Mercury Marine and other major corporations already know, proven effectiveness in reaching target audiences with well designed film or video programs. We are a nation, like it or not, which is conditioned to view programs on film or video screens. We are "sold" on concepts, products _and_ ideas by manipulation of our emotional and sometimes even our rational selves by the verisimilitude of the moving images of light. Using, also proven techniques, some of them time-tested in the cauldron of television advertising, we implant messages in our medium, designed to convince our candidate that his or her preconceptions about our geographic area and our company's products are not only incomplete, but in error. Since we know our target audience, choosing and carefully scripting the most appropriate final form

5

for our program to reach that candidate is the next step in our process.

## Film or Video as the Medium

There are two disparate "looks" to the media of film and video. Television programs and commercials with the intent of selling the concept of "quality" are shot on motion picture film. The look of film is difficult to describe. "Slick", "three-dimensional", "glossy", "professional" are adjectives which have been used in the attempt. Familiar examples are simple to give. Television shows like "Cagney and Lacey", "Hill Street Blues", "M*A*S*H" and all other dramatic series are recorded on film. So are **all** major, national TV spots.

The "video look" is easier to describe. "Live", "one-dimensional" and "bright" are suitable adjectives. Simply compare the "look" of the 6 o'clock news or a Pierquets' TV spot with a Miller, Budweiser, or Ford commercial along with any of the shows named above and you have the idea.

We have the capability to shoot in either medium. Here are the advantages we see for recommending film:

[1] Lighting for film can be more dramatically compelling than the flatter ratios needed for video.

[2] Film imparts a subtle feeling of high quality in the viewer.

[3] A film can be shown to large audiences in theatre or auditorium settings.

[4] Film can be transferred to video with infinitely better results than video can be transferred to film.

6

The negative side of the equation is that it is difficult to intercut scenes shot on film with scenes shot on video. (Not technically, but esthetically). If the final project is going to incorporate a great deal of existing Mercury Marine footage already recorded as video, then our advice would be stay with that medium. The cost factor would be only slightly higher shooting the original on film.

### SECTION TWO

### THE CONCEPT

Crafting a script to tell our story is the heart, soul and backbone of the project. We have two primary considerations:

[1] The ideal target candidate.

[2] Getting the project to that candidate.

Our person is presumed to be intelligent, informed, curious, upwardly mobile, probably an urbanite and also, to face facts, just as probably a man. If the person is a woman in the field, she will no doubt have the same characteristics, however. Because of the type of input we are going to want from this person as a member of the Mercury Marine Team, we are also presuming an inherent sense of adventure and one who enjoys testing his or her limits in a variety of ways. Our major appeal is going to be to someone who feels either bored, boxed-in or frustrated by the limitations of their present position.

While our perfect candidate is represented by a rather narrow spectrum of the entire population, we want to maximize the exposure to most of the potential prospects. Therefore, we propose a standard television length format. There are four from

7

which to choose. A commercial TV one-hour is 48 minutes. A PBS one-hour is 54 minutes. A commercial half-hour is 26 minutes and a PBS show is 28 minutes. It is possible for us to "cut" a program in both the commercial and PBS lengths no matter whether we opt for the hour or half-hour format. The shorter version is more suited for educational classroom, trade/technical conventions and service club showings.

## THE PSYCHOLOGY OF THE SCRIPT

Now that we know our intended viewer and the right format to use, the next step is to construct a script. And here's what our script has to do: **MOTIVATE** the viewer to want to work for Mercury Marine in Wisconsin. Our goal is to initiate a flood of resumes onto the desk of one Neil A. Newman!

Since we anticipate making a film which will be of general enough interest to run as a television program, we are **not** considering a "sales pitch" per se. Even if we were to make this a direct appeal to Mr. and Ms. Engineer, we would not want to appear to be pleading with them. There are two antique axioms which apply to the psychology of our script. "Nothing succeeds like success" and "Everybody loves a winner".

Our script is going to be a positive statement, therefore, about Mercury Marine Product Engineering and about the numerous bounties of the State of Wisconsin.

Our title, **"The Mechanisms of Joy"**, at once sets the thematic statement of our piece and intrigues the viewer to discover what these wonderful mechanisms might be. With the highly visual dynamics at our disposal, we know there will be no

8

difficulty in demonstrating several things about working for Mercury Marine. The sheer fun in our products is demonstrable. We can <u>see</u> them at play in a stunning variety of situations from quiet fishing to world class racing to the <u>adult</u> fantasyland at Disney World; a place which has proven that none of us, and most especially the kind of candidate we're after,  ever **really** "grows-up"!

Our working conditions nurture the perpetual inquisitive "kid" in every one of us. And as for "pushing the edge of the envelope", the present and the future of our marine technology is as fraught with science fiction devices as anything to do with outer space. Our explorations are with "inner space"; the waterways and oceans of Planet Earth and discovering the best and most efficient ways to move in and through them. We see this facet in our gem as central to the theme. We do not build the juggernauts of destruction, nor the prosaic mechanisms of ordinary transportation or commerce. We **build** "the Mechanisms of joy"!

Finally, our product engineers are not always limited to a small part in a vast complex of other parts as engineers frequently are at other corporations. At Mercury Marine, our engineer is limited only by the boundaries of individual imagination.

With this understanding of the psycho/philosophical approach, we are now ready to explore a few considerations for the final shape of the script.

9

## THE ELEMENTS OF THE SCRIPT

We suggest basing a dramatic story around the life of a single Product Engineer at Mercury Marine. We envision having a professional actor play the part. There are several very competent, correct-looking performers available to us in the Fox Valley. We want to build our story around this single central character as he approaches and successfully goes through a critical moment in his career.

We suggest this device based on the principle that viewers can identify with a single person or family with similar lifestyles and insights to their own better than they can absorb large, generalized concepts. We are using one engineer to symbolize and represent the "gestalt" of <u>Mercury</u> <u>Marine</u> <u>Product</u> <u>Engineering</u>.

As a gross example of this dramatic technique, it is impossible for individuals to become terribly involved with or moved by the concept of "slavery". The highly acclaimed and successful television mini-series on that historical event called, "ROOTS", however, focused on the trials and tribulations of one family's history, giving us all the human insight we needed to comprehend their ultimate triumph. The viewer's involvement with our engineer will impart to that viewer the positive message which we have agreed is central to our presentation; Mercury Marine is a great place to work, the Fox Valley of Wisconsin is a great place in which to live.

In working out the plot and the dramatic structure of our script, through a period of time in the life of our central

10

character, we move through four basic stages.

[1] Exposition. This first portion of the script introduces us to the central character, the location, other characters with whom he will interact and sets the stage for the action with which we will become involved.

[2] Point of attack. This is the place in the story where the action begins. It is usually brought on when the central character sets out to achieve a goal.

[3] Rising action. The story picks up pace as the character encounters two dramatic devices called "complication" and "crisis". The complication is some set of circumstances which stand in the way of the character's goal. The crisis is reached when the character has to overcome that complication to achieve the goal.

[4] Resolution and denouement. The resolution of the crisis leads to the third dramatic script device of the "climax" where the character triumphs. Denouement is simply letting the audience see the happy results of the victory in the life of the central character and a place for summing up the message if necessary.

All of these elements follow the plot. Plot means literally "action" and can normally be defined very succinctly. "Boy meets girl, boy loses girl, boy gets girl". Here, for example, is the plot of **HAMLET**: "A young prince has trouble making up his mind whether or not to avenge the death of his father".

The plot is also called "the spine" of the piece. Everything which fleshes out that spine into a full-blown creation with depth and scope hangs on the skeleton of the

11

screenplay.

Attached, part by part, to the spine of our script will be all of the clarifications of the misconceptions about Wisconsin and working on boat motors which we discussed earlier. Those ill conceived notions about Mercury Marine Product Engineers are the easiest to overcome. The freedom of our engineers, the in-depth involvement with each project which an engineer enjoys, the state-of-the-art facilities and support equipment with which he works, and the highly visual, dynamic end-products actually in use are visualizations of the very essence of "the Mechanisms of joy". We can make the computer graphics, the test stations, the metalurgy and so on appear just as exciting visually as anything from the aerospace industry and more. We will demonstrate that our engineer is, indeed, "pushing the envelope" of "high-tech". We also want to show **where** he gets to do so.

We do not intend to make "excuses" for Wisconsin, for we know that none need to be made. The "kernels of truth" which we discussed earlier only need to be shown after they have been planted in rich soil and have grown to mature plants for harvesting in the minds of our viewer. Let's consider those we have agreed upon.

[1] It is true that life in Wisconsin does not follow the pace of urban centers like Los Angeles or New York. But we are far from being a "cultural backwater". The arts and sciences flourish here in cities like Milwaukee and Madison. Both are short hops by either automobile or airplane from our sylvan residence in the Fox Valley. The Valley, itself, is not a "hick

12

town", but rather a slowly expanding urban corridor with fine music, art, theater, institutions of higher education, museums and world class attractions like the E.A.A. Aviation Center and the annual Convention and Fly-in.

Urbanites will be shown a place where middle income people can and do enjoy the goodness of country living, if they choose, on farmettes where no one even thinks of locking doors or worrying about major robbery. Where children can learn about nature by growing up in it. In our small towns we still find cheerful neighbors who care about us and about their community and where, again, we do not have to barricade ourselves behind quadruple, dead-bolted, doors of terrifically overpriced apartments, or fear for ourselves, our wives or our children walking our pleasant, beautifully maintained, tree covered streets night or day. Our values are intact in the midst of the terrors of our urban cousins. We are communities of porch sitters and dog-walkers where the nightly news is filled with feature stories in the main, since there is so little crime to report. Set that notion in the mind of a guy in Houston or Atlanta, show him the red barns and silos of Grandma Moses, the joggers and bicyclists of our autumnally incarnadined countryside and we have set the hook of envy right off the bat!

[2] Yes, Winter is long and harsh. It can also thrill the soul of the poet with its enormous beauty; the crystalline glitter of fresh-fallen snow, the cheerful glow of a house at Christmas looking like a picture postcard, the opportunities for new and adventurous sports like ice-boating, cross-country and

13

downhill skiing minutes away from where we live, snowmobiling,
ice fishing and brisk hikes. There is a grandeur to winter as
part of the visible passing of the seasons of our planet which
many find  endearing and spiritual as we compare and contrast the
coming of spring and summer. Life in our climate presents new
challenges to people who enjoy challenge. And we have already
concluded that it is that very personality whom we are trying to
reach.

For our engineer, Winter also has a flip side with trips to
the Florida facility. We won't deny winter, but we will show  it
for what it is; simply another sparkling facet of existence in
harmony with the spinning of the Earth.

[3] Wisconsinites pay high taxes. Wisconsinites  also enjoy
incredible services in comparision with most other states as a
result. This year, for example, Wisconsin secondary students
scored highest in the nation on the annual college entry
examinations. We have a proud history of education here with a
top-ranked university system which is enormously accessible to
every resident. In the fields of elder care, special education,
and other social services, Wisconsin ranks second to none. And
while we pay for these services, we also enjoy an overall cost of
living which ranks toward the bottom of the national scale for
housing, clothing, food, entertainment and so forth. We may pay
more in property tax on a percentage basis than our friends in
Texas, for example, but we can afford the house we choose to buy!

[4] Our state, does indeed, have a have a blue-collar
element. But that element is a source of pride, not scorn. Our

14

workers are imbued with the "work ethic" as no others in the country. We have craftsmen, not "assemblers". And many of our citizens enjoy the hearty laughter of the local tavern, the bowling alley or the fishing tournament. We are a diverse, conglomeration of ethnic types, taking pride in the heritage left us by the rock-hard pioneers who settled here. We also have a socially elevated class in Wisconsin which enjoys symphony music, art museums and live theatre without the "preciocity" displayed by many urban snobs. It is our cultural intermixing, that freedom of association and our characteristic casualness of dress and manner which lends any Wisconsinite a sense of belonging, of having roots and of loving this land of rolling fields, forests, streams, lakes and rivers.

And, in the experience of our central character, we will blend all of these elements of the script and of the story together to produce a revelry of sights and sounds which will leave all but the most blase yearning to become a part of it all.

### AT YOUR SERVICE

Any film begins with the written word. We are ready to begin scripting **The Mechanisms of Joy.** The elements are all waiting to be assembled. We propose the following schedule to implement the project.

[A] Research and scriptwriting: October 14th through November 14th, 1985.

[B] Production: November 18th through May 1st, 1986 to enable us to show the transition of seasons.

[C] Post production: May 1st through June 1st, 1986.

15

## BUDGET

It is impossible to present a detailed production budget since the amounts will depend upon the exigencies of the script from which the actual budget must be drawn. We can, however, provide the actual cost of research and scriptwriting and a "ballpark" figure between which the production budget will fall for either shooting on original film or original video.(The top figure represents an estimated minimum, the bottom an estimated maximum).

    [A] Research and Scriptwriting..............$ 1,200.00

    [B] Production for Film....................$ 5,000.00
                                                20,000.00

    [C] Production for Video..................$ 4,000.00
                                                20,000.00

    [D] Post-production for Film...............$ 3,000.00
                                                 7,000.00

    [E] Post-production for Video.............$ 2,500.00
                                                 7,000.00

Mercury Marine shall own all rights to the script upon completion and approval. At that time we will present a detailed, accurate production budget estimate for your consideration. We invite you to solicit comparison bids on the production at that time.

## TERMS

    [A] Research and Scriptwriting: $600.00 on contract signing, $600.00 balance on script approval by the client.

    [B] Production and Post-Production: One half on contract

16

signing, one half balance due on delivery of final print.

## CONCLUSION

We believe that you have a challenging and worthwhile project, the goals for which are achievable by us at the highest quality and the lowest cost. Our bid for this project will be on a cost recovery basis, paying only for the actual materials and services needed. We want to prove to you that Mercury Marine does not need to go outside the Fox Valley to provide itself with the best film and video production services available. We regard this as our chance to do so. A copy of some of our previous credits is attached.

We are excited at the prospect of working on this entertaining and stimulating project and look forward to a long and happy association with you.

# Selected Bibliography

## ACCOUNTING AND BOOKKEEPING

Dixon, Robert L. *The Executive's Accounting Primer*. New York: McGraw-Hill, 1971.

*How to Read a Financial Report*. New York: Merrill Lynch, Pierce, Fenner & Smith, Inc., 1971

Katz, Benjamin. *The Why, What and How of Recordkeeping*. New York: Overlook Press, 1973.

## ADVERTISING AND MARKETING

Dunn, S. Watson. *Advertising*. New York: Holt, Rinehart & Winston, 1969.

Jewler, A. Jerome. *Creative Strategy in Advertising*. Belmont, CA: Wadsworth Inc., 1981.

McCarthy, E.J. *Basic Marketing*. Homewood, IL: Richard D. Irwin, Inc., 1971.

Moriarty, Sandra E. *Creative Advertising: Theory and Practice*. Englewood Cliffs, NJ: Prentice-Hall, 1986.

Russell, Thomas, and Verrill, Glenn. *Otto Kleppner's Advertising Procedure*. 9th ed. Englewood Cliffs, NJ: Prentice-Hall, 1986.

Warner, Charles. *Broadcast and Cable Selling*. Belmont, CA: Wadsworth Inc., 1986.

## CAPITAL

Belew, Richard. *How to Win Profits and Influence Bankers*. New York: Van Nostrand Reinhold Co., 1973.

*Building Strong Relations with Your Bank*. SBA Publication No. 107. Fort Worth, TX: Small Business Administration.

*What to Do if You Want to Apply for a Loan from the Small Business Administration*. Fort Worth, TX: Small Business Administration.

## CREDIT POLICIES

Blake, William Henry. *Retail Credit and Collections*. SBA Bibliography No. 31. Fort Worth, TX: Small Business Administration.

Kolodny, Leonard. *Outwitting Bad Check Passers*. SBA Publication No. 137. Fort Worth, TX: Small Business Administration.

## GENERAL BUSINESS KNOWLEDGE

Bittel, Lester. *The Nine Master Keys of Management*. New York: McGraw-Hill, 1982.

*The Pitfalls of Business*. New York: Dun & Bradstreet, 1973.

Shelly & Cashman. *Introduction to Computers and Data Processing*. Fullerton, CA: Anaheim Publishing Co., 1980.

Uris, Auren. *The Executive Deskbook*. New York: Van Nostrand Reinhold Co., 1970.

## GOVERNMENT FORMS AND REPORTS

*Your Business Tax Kit*. Washington, DC: Internal Revenue Service. (Published annually.)

## LOCATING A BUSINESS

Epplin, Rose, and Roussel, F.J. *Thinking About Going Into Business?* SBA Management Aid No. 2.025. Fort Worth, TX: Small Business Administration.

Lowry, James R. *Using a Traffic Study to Select a Retail Site*. SBA Publication No. 152. Fort Worth, TX: Small Business Administration.

Weber, Fred, Jr. *Locating or Relocating Your Business*. SBA Management Aid No. 201. Fort Worth, TX: Small Business Administration.

## MANAGEMENT KNOWLEDGE

Drucker, Peter. *The Practice of Management*. rev. ed. New York: Harper & Row, 1980.

Klall, Lawrence. *Managing the Dynamic Small Firm*. Belmont, CA: Wadsworth Inc., 1971.

Metcalf, Wendell O. *Starting and Managing a Small Business of Your Own*. Fort Worth, TX: Small Business Administration, 1973.

## PERSONNEL

*Productivity—The Personnel Challenge*. Englewood Cliffs, NJ: Prentice-Hall, 1973.

Rabe, William F. *Matching the Applicant to the Job*. SBA Management Aid No. 185. Fort Worth, TX: Small Business Administration.

Raphelson, Rudolph. *Finding and Hiring the Right Employee*. SBA Management Aid No. 106. Fort Worth, TX: Small Business Administration.

Smith, Leonard. *Checklist for Developing a Training Program*. SBA Management Aid No. 106. Fort Worth, TX: Small Business Administration.

## TIME MANAGEMENT

Wantola, Stanley. *Delegating Work and Responsibility*. SBA Management Aid No. 191. Fort Worth, TX: Small Business Administration.

## VIDEO

Anderson, Gary H. *Video Editing and Post-Production: A Professional Guide*. White Plains, NY: Knowledge Industry Publications, Inc., 1984.

Blythin, Evan, and Samovar, Larry A. *Communicating Effectively on TV*. Belmont, CA: Wadsworth Inc., 1985.

Cremer, Charles F., and Yoakam, Richard D. *Television News and New Technology*. New York: Random House, 1985.

Efrein, Joel Lawrence. *Videotape Production & Communication Techniques*. Blue Ridge Summit, PA: TAB Books, 1979.

Hiebert, Ungurait & Bohn. *Mass Media II*. New York: Longman Inc., 1979.

Kelly, Eugene W., Jr. *Effective Interpersonal Communication: A Manual for Skill Development*. Lanham, MD: University Press of America, 1985.

*Lighting Handbook for Television, Theatre and Professional Photography*. 7th ed. Danvers, MA: Sylvania/GTE Products Corp., 1984.

Mathias, Harry, and Patterson, Richard. *Electronic Cinematography*. Belmont, CA: Wadsworth Inc., 1985.

McCavitt, William E. *Television Technology: Alternative Communication Systems*. Lanham, MD: University Press of America, 1985.

Millerson, Gerald. *TV Lighting Methods*. New York: Hastings House, 1979.

Mintz, Harold K., and Vasile, Albert J. *Speak With Confidence: A Practical Guide*. 2d. ed. Cambridge, MA: Winthrop Publishers, 1980.

## WRITING / PRODUCING

Brenner, Alfred. *The TV Scriptwriter's Handbook*. Cincinnati, OH: Writer's Digest Books, 1985.

Clark, Bernadine, ed. *Writer's Resource Guide*. 2d ed. Cincinnati, OH: Writer's Digest Books, 1985.

Field, Syd. *Screenplay: The Foundations of Screenwriting*. New York: Delta Books, 1982.

Polking, Kirk, ed. *Jobs for Writers*. Cincinnati, OH: Writer's Digest Books, 1985.

Polking, Kirk, ed. *Writer's Encyclopedia*. Cincinnati, OH: Writer's Digest Books, 1985.

Polking, Kirk, and Meranus, Leonard S., eds. *Law and the Writer*. 3d. ed. Cincinnati, OH: Writer's Digest Books, 1985.

Straczynski, J. Michael. *The Complete Book of Scriptwriting*. Cincinnati, OH: Writer's Digest Books, 1985.

Vale, Eugene. *The Technique of Screen and Television Writing*. rev. ed. Englewood Cliffs, NJ: Prentice-Hall, 1982.

# Index

## ABOUT THE AUTHOR

Dr. Bob Jacobs is a professor of Radio-TV-Film at the University of Wisconsin-Oshkosh. He also owns and operates an independent production company and is a prolific free-lance writer. He has published articles and book chapters for *Writer's Digest Books,* has written articles for *American Cinemeditor, Today's Filmmaker, American Film, Speech Tech, Telek* and *Video Manager.* He recently completed his first novel, *Season of the Beast.*

Before turning to teaching, Jacobs had a varied career as a free-lance writer/director in his native Hollywood. His credits there include the "True-life Adventure" television series for Walt Disney Productions, "The Midnight Special," "The Dionne Warwick Special," "Beethoven, Bach and The Beatles," "The Great American Dream Machine" and hundreds of radio and television spot commercials.

As an independent producer, Jacobs offers his students at UW-Oshkosh a taste of the "real world" by having them work on projects for his company, The Film Farm. In addition to industrial and television commercial projects, Jacobs has been involved in the production of two feature-length movies in Oshkosh: *Exit Dying,* starring Henry Darrow and *Dreams Come True.* He is currently at work on a series of educational films and videos narrated by William Conrad for distribution by Encyclopaedia Britannica.

Jacobs holds a B.A. in Cinema from the University of Southern California and a Ph.D. in Dramatic Art from the University of California-Santa Barbara. He has lectured on independent production at the University of Texas, the University of Minnesota, Purdue University, California State University-Long Beach, California State University-Humboldt, Milwaukee Area Technical College and the University of Wisconsin-Milwaukee. He has also acted as a consultant to a number of corporations and independent production companies throughout the country.